Holt Mathematics

Chapter 6 Resource Book

HOLT, RINEHART AND WINSTON
A Harcourt Education Company

Orlando • Austin • New York • San Diego • London

Copyright © by Holt, Rinehart and Winston

All rights reserved. No part of this publication may be reproduced or transmitted in any form or by any means, electronic or mechanical, including photocopy, recording, or any information storage and retrieval system, without permission in writing from the publisher.

Teachers using HOLT MATHEMATICS may photocopy complete pages in sufficient quantities for classroom use only and not for resale.

Printed in the United States of America

If you have received these materials as examination copies free of charge, Holt, Rinehart and Winston retains title to the materials and they may not be resold. Resale of examination copies is strictly prohibited and is illegal.

Possession of this publication in print format does not entitle users to convert this publication, or any portion of it, into electronic format.

ISBN 0-03-078396-8

CONTENTS

Blackline Masters

Parent Letter	1
Lesson 6-1 Practice A, B, C	3
Lesson 6-1 Reteach	6
Lesson 6-1 Challenge	7
Lesson 6-1 Problem Solving	8
Lesson 6-1 Reading Strategies	9
Lesson 6-1 Puzzles, Twisters & Teasers	10
Lesson 6-2 Practice A, B, C	11
Lesson 6-2 Reteach	14
Lesson 6-2 Challenge	16
Lesson 6-2 Problem Solving	17
Lesson 6-2 Reading Strategies	18
Lesson 6-2 Puzzles, Twisters & Teasers	19
Lesson 6-3 Practice A, B, C	20
Lesson 6-3 Reteach	23
Lesson 6-3 Challenge	25
Lesson 6-3 Problem Solving	26
Lesson 6-3 Reading Strategies	27
Lesson 6-3 Puzzles, Twisters & Teasers	28
Lesson 6-4 Practice A, B, C	29
Lesson 6-4 Reteach	32
Lesson 6-4 Challenge	33
Lesson 6-4 Problem Solving	34
Lesson 6-4 Reading Strategies	35
Lesson 6-4 Puzzles, Twisters, & Teasers	36
Lesson 6-5 Practice A, B, C	37
Lesson 6-5 Reteach	40
Lesson 6-5 Challenge	42
Lesson 6-5 Problem Solving	43
Lesson 6-5 Reading Strategies	44
Lesson 6-5 Puzzles, Twisters & Teasers	45
Lesson 6-6 Practice A, B, C	46
Lesson 6-6 Reteach	49
Lesson 6-6 Challenge	51
Lesson 6-6 Problem Solving	52
Lesson 6-6 Reading Strategies	53
Lesson 6-6 Puzzles, Twisters & Teasers	54
Lesson 6-7 Practice A, B, C	55
Lesson 6-7 Reteach	58
Lesson 6-7 Challenge	60
Lesson 6-7 Problem Solving	61
Lesson 6-7 Reading Strategies	62
Lesson 6-7 Puzzles, Twisters & Teasers	63
Answers to Blackline Masters	64

Date _____

Dear Family,

In this chapter, your child will learn to work with percents: how to relate decimals, fractions, and percents; how to find percent of increase and decrease; and various applications using percents. Your child will relate percent to the tests in school, sales at the mall, commission, and loans.

Percents are ratios that compare a number to another number. Your child will learn how to relate decimals, fractions, and percents.

To convert a fraction to a decimal, divide the numerator by the denominator as shown.

$\frac{1}{8} = 1 \div 8 = 0.125$

To convert a decimal to a percent, multiply by 100 and insert the percent symbol.

$0.125 \cdot 100 = 12.5\%$

Next, your child will learn to work with the three types of percent problems.

Three Types of Percent Problems	
1. Finding the percent of a number	15% of 120 = n
2. Finding the percent one number is of another	p% of 120 = 18
3. Finding a number when the percent is known	15% of n = 18

Your child will learn how to work with percents in various applications, such as the following exercise involving a **commission rate.**

You can multiply by percents to find the commission amounts.

A sales person sold a car for $39,500, earning a 4% commission on the sale. How much was her commission?

Think: commission rate • sales = commission.

4% • $39,500 = c *Write an equation.*

0.04 • 39,500 = c *Change the percent to a decimal: 4% = $\frac{4}{100}$ = 0.04.*

1580 = c *Solve for c.*

She earned a commission of $1580 on the sale.

Holt Mathematics

Your child will also use an equation to a number when the percent is known.

42 is 5% of what number?

$42 = 5\% \cdot n$	*Set up an equation.*
$42 = 0.05n$	*Change the percent to a decimal.*
$\dfrac{42}{0.05} = \dfrac{0.05}{0.05}n$	*Divide both sides by 0.05.*
$840 = n$	

42 is 5% of 840.

Another application involving percent is computing simple interest.

> To find **simple interest** I, use the formula $I = Prt$, where
>
> $P =$ **principal,** the amount of money borrowed or invested,
>
> $r =$ **rate of interest,** the percent charged or earned, and
>
> $t =$ **time** that the money is borrowed or invested (in years).

Interest = Principal · rate · time $I = Prt$

Tristan borrowed $14,500 from his brother and promised to pay him back over 5 years at an annual simple interest rate of 7%. How much interest will he pay if he pays off the entire loan at the end of the fifth year?

$I = P \cdot r \cdot t$	*Use the formula.*
$I = 14{,}500 \cdot 0.07 \cdot 5$	*Substitute. Use 0.07 for 7%.*
$I = 5075$	*Solve for I.*

Tristan will pay $5075 in interest.

You can find the total amount A to be repaid on a loan by adding the principal P to the interest I.

Principal + interest = amount $P + I = A$

What is the total amount Tristan will repay?

$P + I = A$	*Use the formula.*
$14{,}500 + 5075 = A$	*Substitute.*
$19{,}575 = A$	*Solve for A.*

Tristan will repay a total of $19,575 on his loan.

For additional resources, visit go.hrw.com and enter the keyword MT7 Parent.

Name _____ Date _____ Class _____

LESSON 6-1 Practice A
Relating Decimals, Fractions, and Percents

Find the missing ratio or percent equivalent for each letter on the number line.

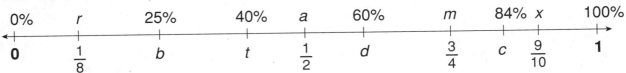

1. a
2. b
3. c
4. d

5. m
6. r
7. t
8. x

Compare. Write <, >, or =.

9. $\frac{1}{2}$ ☐ 20%
10. 60% ☐ $\frac{4}{5}$
11. 37% ☐ 0.37
12. 0.76 ☐ 81%
13. $\frac{9}{10}$ ☐ 99%
14. 0.7 ☐ 7%

Order the numbers from least to greatest.

15. $\frac{3}{4}$, 0.4, 0.34, 45%
16. $\frac{2}{3}$, 30%, 0.03, 23%

17. 100%, 0.95, 59%, $\frac{1}{9}$
18. 62%, $\frac{1}{2}$, 0.58, 85%

19. Katrina spent $\frac{1}{2}$ of the money she received for her birthday on a new sweater. What percent of the money she received did she spend on the sweater? _____

20. Mr. Laschat has 20 students in his class. If 12 of them are girls, what percent of the class are girls? _____

Name _____ Date _____ Class _____

LESSON 6-1 Practice B
Relating Decimals, Fractions, and Percents

Find the missing ratio or percent equivalent for each letter on the number line.

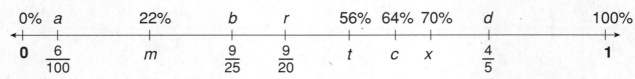

1. a

2. b

3. c

4. d

_____ _____ _____ _____

5. m

6. r

7. t

8. x

_____ _____ _____ _____

Compare. Write <, >, or =.

9. $\dfrac{3}{4}$ ☐ 70%

10. 60% ☐ $\dfrac{3}{5}$

11. 58% ☐ 0.6

12. 0.09 ☐ 15%

13. $\dfrac{2}{3}$ ☐ 59%

14. 0.45 ☐ 40.5%

Order the numbers from least to greatest.

15. 99%, 0.95, $\dfrac{5}{9}$, 9.5%

16. $\dfrac{3}{8}$, 50%, 0.35, 38%

_____ _____

17. $\dfrac{4}{5}$, 54%, 0.45, 44.5%

18. $\dfrac{1}{3}$, 20%, 0.3, 3%

_____ _____

19. There are 25 students in math class. Yesterday, 6 students were absent. What percent of the students were absent? _____

20. Albert spends 2 hours a day on his homework and an hour playing video games. What percent of the day is this? _____

21. Ragu ran the first 3 miles of a 5 mile race in 24 minutes. What percent of the race has he run? _____

Name _____ Date _____ Class _____

LESSON 6-1 Practice C
Relating Decimals, Fractions, and Percents

Compare. Write <, >, or =.

1. $\dfrac{1}{3}$ ☐ 32%
2. 87.5% ☐ $\dfrac{7}{8}$
3. 99% ☐ 1

Order the numbers from least to greatest.

4. 98%, 0.94, $\dfrac{5}{4}$, 95.9%

5. $\dfrac{1}{5}$, 5%, 0.35, 3.5%

_____

_____

6. $\dfrac{5}{8}$, 65%, 0.5, 60.5%

7. $\dfrac{2}{3}$, 60%, 0.06, 6.5%

_____

_____

Write the labels from each circle graph as percents.

8. 0.6 = _____ %
   0.1 = _____ %
   0.3 = _____ %

9. $\dfrac{1}{6}$ = _____ %
   $\dfrac{1}{4}$ = _____ %
   $\dfrac{7}{12}$ = _____ %

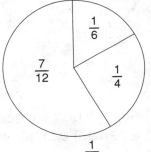

10. 0.06 = _____ %
    0.18 = _____ %
    0.41 = _____ %
    $\dfrac{7}{20}$ = _____ %

11. $\dfrac{1}{20}$ = _____ %
    $\dfrac{1}{4}$ = _____ %
    $\dfrac{11}{50}$ = _____ %
    $\dfrac{12}{25}$ = _____ %

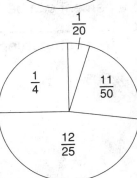

12. Of the 150 students in the eighth grade, 6 were absent yesterday. What percent of eighth graders were absent yesterday? _____

13. A student has read 23 out of 41 pages assigned for homework. What percent of the pages has the student read? _____

14. Tish has 17 relatives that live out of state and 12 that live in state. What percent of her relatives live in state? _____

15. A molecule of nitric acid is made up of 3 molecules of oxygen, 1 molecule of hydrogen, and 1 molecule of nitrogen. What percent of the atoms of nitric acid is oxygen? _____

Reteach
6-1 Relating Decimals, Fractions, and Percents

A **percent** (symbol %) is a *ratio*, where the comparison is to the number 100.

The ratio is then written in simplest form.

$$40\% = \frac{40}{100} = \frac{40 \div 20}{100 \div 20} = \frac{2}{5}$$

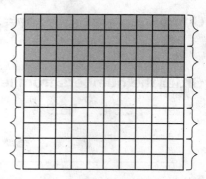

Write each percent as a ratio in simplest form.

1. $80\% = \frac{}{100}$

   $= \frac{ \div }{100 \div 20}$

   $= \underline{}$

2. $37.5\% = \frac{}{100}$

   $= \frac{}{1000}$

   $= \frac{ \div }{1000 \div 125}$

   $= \underline{}$

3. $65\% = \frac{}{100}$

   $= \frac{ \div }{100 \div 5}$

   $= \underline{}$

Since a percent compares a number to 100, a percent can be written as a decimal. $40\% = \frac{40}{100} = 0.40$

Write each percent as a decimal.

4. $80\% = \frac{}{100}$

   $= \underline{}$

5. $37.5\% = \frac{}{100}$

   $= \frac{}{1000}$

   $= \underline{}$

6. $65\% = \frac{}{100}$

   $= \underline{}$

Use the results of Exercises 1–6 to compare. Write <, >, or =.

7. $\frac{13}{20}$ ☐ 37.5%

8. 80% ☐ 0.65

9. 65% ☐ 0.8

10. 0.375 ☐ 80%

11. 37.5% ☐ 0.65

12. 65% ☐ $\frac{13}{20}$

Name _____ Date _____ Class _____

  Challenge
6-1 *100% Filled*

Materials needed: colored pencils or pens

For each exercise, select from the box a different combination of numbers whose sum is 100%. An item may be used only once in a combination, but may be used again in a different combination. Write your selection on the line below the grid. Shade the squares in the grid with a different color for each number you selected.

68%	$\frac{1}{2}$	60%	$\frac{1}{20}$	1%	0.9	0.04	$\frac{13}{20}$	$\frac{1}{7}$
$\frac{3}{25}$	16%	55%	0.02	$\frac{1}{9}$	0.15	$\frac{1}{8}$	24%	0.44
0.06	$\frac{1}{6}$	$\frac{1}{4}$	29%	0.037	$\frac{1}{3}$	19%	$\frac{17}{50}$	

1.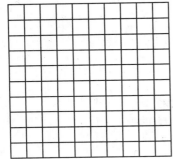

   _____

2.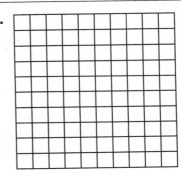

   _____

3.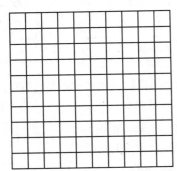

   _____

Holt Mathematics

Name _____ Date _____ Class _____

LESSON 6-1 Problem Solving
Relating Decimals, Fractions, and Percents

The table shows the ratio of brain weight to body size in different animals. Use the table for Exercises 1–3. Write the correct answer.

1. Complete the table to show the percent of each animal's body weight that is brain weight. Round to the nearest hundredth.

2. Which animal has a greater brain weight to body size ratio, a dog or an elephant?

   _____

3. List the animals from least to greatest brain weight to body size ratio.

   _____

Animal	Brain Weight / Body Weight	Percent
Mouse	$\frac{1}{40}$	
Cat	$\frac{1}{100}$	
Dog	$\frac{1}{125}$	
Horse	$\frac{1}{600}$	
Elephant	$\frac{1}{560}$	

The table shows the number of wins and losses of the top teams in the National Football Conference from 2004. Choose the letter of the best answer. Round to the nearest tenth.

4. What percent of games did the Green Bay Packers win?

   A 10% C 37.5%
   B 60% D 62.5%

Team	Wins	Losses
Philadelphia Eagles	13	3
Green Bay Packers	10	6
Atlanta Falcons	11	5
Seattle Seahawks	9	7

5. Which decimal is equivalent to the percent of games the Seattle Seahawks won?

   F 0.05625 H 5.625
   G 0.5625   J 56.25

6. Which team listed had the highest percentage of wins?

   A Philadelphia Eagles
   B Green Bay Packers
   C Atlanta Falcons
   D Seattle Seahawks

Copyright © by Holt, Rinehart and Winston.
All rights reserved.

Holt Mathematics

Name _____ Date _____ Class _____

Reading Strategies
LESSON 6-1 Multiple Representations

Three out of four middle school students have homework each night. Three out of four is a ratio and can be shown in different ways.

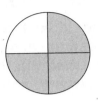

The ratio can be written as a fraction. → $\frac{3}{4}$

The ratio can be written as a decimal. → 0.75
Divide the numerator by the denominator to get the equivalent decimal.
3 ÷ 4 = 0.75

The ratio can be read as a fraction with a denominator of 100.   Read: → "75 hundredths"

The ratio can be written as a fraction with a denominator of 100. → $\frac{75}{100}$

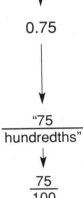

Percent means "per hundred." $\frac{75}{100}$ can be written as 75%.

Use this statement to complete Exercises 1–6:
One out of five students has a younger brother.

1. Write this ratio as a fraction.
2. How can you find the decimal that is equal to $\frac{1}{5}$?
3. Write the decimal you get when you divide 1 by 5.
4. What does *percent* mean?
5. Write a fraction equal to $\frac{1}{5}$ and 0.2 that has a denominator of 100.
6. Write that fraction as a percent.

Name _____ Date _____ Class _____

LESSON 6-1 Puzzles, Twisters & Teasers
A Slice of Soccer Fun!

To find the answer to the riddle, label the unidentified portion of each circle graph. Express the answers as decimals. Each decimal is paired with a letter. Put the letter that matches the decimal in the puzzle answer to solve the riddle.

1. N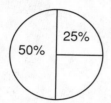

2. Y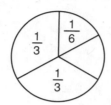

3. E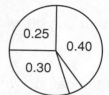

4. S

5. E

6. E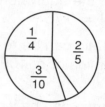

7. W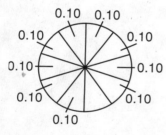

8. J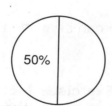

9. R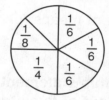

Where is the best place to buy a new soccer shirt?

___ ___ ___   ___ ___ ___ ___ ___ ___
0.25 0.05 0.10   0.50 0.05 0.125 0.45 0.05 0.17

Name _____ Date _____ Class _____

LESSON 6-2 Practice A
Estimate with Percents

Write a benchmark you could use to estimate each percent.

1. 9% _____   2. 73% _____   3. 26% _____

4. 48% _____   5. 19% _____   6. 53% _____

7. 34% _____   8. 12% _____   9. 65% _____

Estimate.

10. 10% of 98 _____   11. 48% of 83 _____   12. 20% of 42 _____

13. 40 out of 125 _____   14. 33% of 45 _____   15. 11 out of 215 _____

16. 25% of 33 _____   17. 21% of 49 _____   18. 30% of 36 _____

19. Felicia spent $8.78 of her $10.00 weekly allowance. About what percent of her allowance did she spend? _____

20. A restaurant bill for dinner is a total of $49.60. Estimate the amount to leave as a 20% tip. _____

21. A company has found that on average about 8% of the light bulbs they manufacture are defective. Out of 998 bulbs, the manager assumes that about 80 are defective. Is the manager's estimate reasonable? Explain. _____

Holt Mathematics

Name _____ Date _____ Class _____

LESSON 6-2 Practice B
Estimate with Percents

Estimate.

1. 74% of 99

2. 25% of 39

3. 52% of 10

_____ _____ _____

4. 21% of 50

5. 30% of 61

6. 24% of 48

_____ _____ _____

7. 5% of 41

8. 50% of 178

9. 33% out of 62

_____ _____ _____

Estimate.

10. 48% of 30 is about what number?

11. 26% of 36 is about what number?

_____ _____

12. 30% of 22 is about what number?

13. 21% of 63 is about what number?

_____ _____

14. Rodney's weekly gross pay is $91. He must pay about 32% in taxes and deductions. Estimate Rodney's weekly take-home pay after deductions. _____

15. In the last school election, 492 students voted. Mary received 48% of the votes. About how many votes did she receive? _____

16. A restaurant bill for lunch is $14.10. Grace wants to leave a 15% tip and the sales tax rate is 5.5%. About how much will lunch cost Grace in all? _____

17. A company has found that on average about 6% of the batteries they manufacture are defective. Out of 1,385 batteries, the supervisor assumes that about 83 are defective. Estimate to determine if the manager's number is reasonable? Explain. _____

_____

Name _____ Date _____ Class _____

Practice C
LESSON 6-2 *Estimate with Percents*

Estimate each number of percent.

1. 74% of 59 is about what number? _____

2. 32% of 449 is about what number? _____

3. 19% of 60 is about what number? _____

4. 31% of 67 is about what number? _____

5. 9.8% of 710 is about what number? _____

6. 103% of 42 is about what number? _____

7. 39% of 78 is about what number? _____

8. 90% of 388 is about what number? _____

9. 66% of 24 is about what number? _____

10. 18.7% of 98 is about what number? _____

11. 50% of 18.25 is about what number? _____

12. $33\frac{1}{3}$% of 62 is about what number? _____

13. A stock on the New York State Stock Exchange opens at $45.02 and closes up 10% for the day. About how much does stock price increase? _____

14. A driver delivered 462 boxes to a manufacturing company. She unloaded 75% of the boxes before lunch. About how many boxes did she unload? _____

15. A bill for dinner is $35.75. Josh wants to leave a 15% tip. The bill will also include sales tax at a rate of 6.2%. About how much will dinner cost Josh in all? _____

16. On a weekday, about 39% of commuter cars have a single occupant. Out of 2,385 commuter cars, the traffic control officer assumes that about 930 cars have a single occupant. Estimate to determine if the traffic officer's number is reasonable? Explain. _____

Name _____ Date _____ Class _____

LESSON 6-2 Reteach
Estimate with Percents

You can estimate the solutions to different types of problems involving percents by rounding numbers.

Type 1: Finding a percent of a number.

Estimate 38% of 470.

First, round the percent to a common percent with an easy fractional equivalent.

$38\% \approx 40\% = \frac{40}{100} = \frac{2}{5}$

Then, round the number to a number divisible by the denominator of the fraction. | 500 is divisible by 5.

$470 \approx 500$

Use mental math to multiply.

$\frac{2}{5} \cdot 500$ Think: If $500 \div 5 = 100$, then $100 \cdot 2 = 200$

So, 38% of 470 is about 200.

Complete to estimate each percent.

1. 32% of 872

   Round to a percent with an easy fractional equivalent: $32\% \approx$ _____ $\% = \frac{1}{3}$

   Round to hundreds place, divisible by 3: $872 \approx$ _____

   Do mental math to multiply: $\frac{1}{3} \times$ _____ = _____

   So, 32% of 872 is about _____.

2. 78% of 495

   % to easy fraction:

   $78\% \approx$ _____ $\% =$ _____

   divisible by denominator:
   $495 \approx$ _____

   Do mental math to multiply.

   _____ × _____ = _____

   So, 78% of 495 is about _____.

3. 73% of 1175

   % to easy fraction:

   $73\% \approx$ _____ $\% =$

   divisible by denominator:
   $1175 \approx$ _____

   Do mental math to multiply.

   _____ × _____ = _____

   So, 73% of 1175 is about _____.

Name _____ Date _____ Class _____

Reteach
LESSON 6-2 Estimate with Percents (continued)

Type 2: Finding what percent one number is of another.

About what percent of 82 is 39?

Round to numbers with common factors, **compatible numbers**.
$82 \approx 80$ $39 \approx 40$

Write a fraction in the form $\frac{is \text{ number}}{of \text{ number}}$: $\frac{40}{80}$.

Reduce the fraction and convert to percent.

$\frac{40}{80} = \frac{1}{2}$ You may recognize the % now.

$= \frac{1 \cdot 50}{2 \cdot 50} = \frac{50}{100}$ Or, get a fraction with denominator 100.

$= 50\%$

So, 39 is about 50% of 82.

Estimate what percent one number is of another.

4. About what percent of 98 is 27?

   Use compatible numbers. $98 \approx$ _____ $27 \approx$ _____

   Write fraction. $\frac{is \text{ number}}{of \text{ number}}$: _____

   Reduce the fraction and convert to percent. ___ = ___ = ___%

   So, 27 is about _____% of 98.

5. About what percent of 19 is 6?

   $19 \approx$ _____ $6 \approx$ _____

   $\frac{is \text{ number}}{of \text{ number}}$: ___ = ___ = ___%

   So, 6 is about _____% of 19.

6. About what percent of 51 is 89?

   $51 \approx$ _____ $89 \approx 90$

   $\frac{is \text{ number}}{of \text{ number}}$ ___ : $= \frac{9}{5} = 1 +$ _____

   $=$ _____%

   So, 89 is about _____% of 51.

Name _____ Date _____ Class _____

LESSON 6-2 Challenge
As the Wheel Turns

For each wheel, write a percent problem using estimation. In your problem, use the percent at the center of the wheel and two other numbers on the wheel. The first one in each row is done.

1.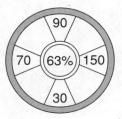

   63% of 150 is about 90.

2.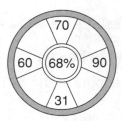

   _____

   _____

3.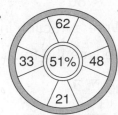

   _____

   _____

4.

   16 is about 23% of 71.

5.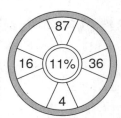

   _____

   _____

6.

   _____

   _____

7.

   141% of 85 is about 120.

8.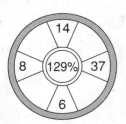

   _____

   _____

9.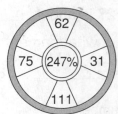

   _____

   _____

16

Holt Mathematics

Name _____ Date _____ Class _____

Problem Solving
LESSON 6-2 Estimate with Percents

Write an estimate.

1. A store requires you to pay 15% up front on special orders. If you plan to special order items worth $74.86, estimate how much you will have to pay up front.

2. A store is offering 25% off of everything in the store. Estimate how much you will save on a jacket that is normally $58.99.

3. A certain kind of investment requires that you pay a 10% penalty on any money you remove from the investment in the first 7 years. If you take $228 out of the investment, estimate how much of a penalty you will have to pay.

4. John notices that about 18% of the earnings from his job go to taxes. If he works 14 hours at $6.25 an hour, about how much of his check will go for taxes?

Choose the letter for the best estimate.

5. In its first week, an infant can lose up to 10% of its body weight in the first few days of life. Which is a good estimate of how many ounces a 5 lb 13 oz baby might lose in the first week of life?
   A 0.6 oz
   B 9 oz
   C 18 oz
   D 22 oz

6. A CD on sale costs $12.89. Sales tax is 4.75%. Which is the best estimate of the total cost of the CD?
   F $13.30
   G $13.55
   H $14.20
   J $14.50

7. In a class election, Pedro received 52% of the votes. There were 274 students who voted in the election. Which is the best estimate of the number of students who voted for Pedro?
   A 70 students
   B 100 students
   C 125 students
   D 140 students

8. Mel's family went out for breakfast. The bill was $24.25 plus 5.2% sales tax. Mel wants to leave a 20% tip. Which is the best estimate of the total bill?
   F $25.45
   G $29.25
   H $30.25
   J $32.25

Holt Mathematics

Name _____ Date _____ Class _____

LESSON 6-2 Reading Strategies
Use Context

An **estimate** is useful when an exact answer is not needed. You can

---

Chris made 15 out of 32 free throws. About what fraction of his free throws did he make?

$\frac{15}{32}$ stands for the number of free throws Chris made out of the total number of baskets shot.

30 is close to 32.

$\frac{15}{30} = \frac{1}{2}$

Chris made about half of his free throws.

There are 36 students in the eighth grade. 28% of them play basketball. **About** how many students play basketball?

25% is close to 28% and much easier to work with.

25% = one-fourth

One-fourth of 36 = 9.

---

use numbers that work well together when you estimate.

Use the following problem for Exercises 1–4:
27.5% of the 40 students on the field trip rode the bus. About how many students rode the bus?

1. Do 27.5% and 40 work well together?

   _____

2. What percent would work better with 40? Explain why.

   _____

3. Rewrite the problem using numbers that work well together.

   _____

4. About how many students rode the bus?

Puzzles, Twisters & Teasers
6-2 Dressed to Fill!

Decide if the statements are true or false. After you have completed all of the exercises take the letters from all of the true statements and unscramble them to solve the riddle.

1. 50% of 297 is about 150   G   __true__
2. 60% of 800 is about 300   M   __false__
3. 50% of 795 is about 200   A   __false__
4. 24% of 78 is about 20   N   __true__
5. 26% of 98 is about 75   W   __false__
6. 25% of 200 is about 100   Y   __false__
7. 31% of 42 is about 13   I   __true__
8. 25% of 925 is about 230   S   __true__
9. 88% of 180 is about 160   E   __true__
10. 15% of 600 is about 150   B   __false__
11. 105% of 776 is about 815   L   __true__
12. 11% of 61 is about 7   R   __true__
13. 55% of 810 is about 450   D   __true__

Why did the tomato blush?

Because it saw the

__S__ A __L__ A __D__ __D__ __R__ E S S __I__ __N__ __G__

Name _____ Date _____ Class _____

LESSON 6-3
Practice A
Finding Percents

Write an equation or proportion that you can use to solve each problem.

1. What percent of 14 is 7?

2. 8 is what percent of 25?

Find each percent.

3. What percent of 60 is 15?

4. 11 is what percent of 44?

5. What percent of 35 is 7?

6. What percent of 90 is 9?

7. 6 is what percent of 40?

8. What percent of 8 is 6?

9. What percent of 50 is 15?

10. 14 is what percent of 56?

11. 9 is what percent of 36?

12. 10 is what percent of 25?

13. What percent of 50 is 12?

14. What percent of 24 is 3?

15. A survey asked 180 people if they like winter, and 45 people said yes. What percent of the people surveyed like winter? _____

16. The Reed family bought a case of apples. Paula ate $\frac{3}{10}$ of the apples. Kyle ate 21% of the apples. Wendy ate 0.34 of the apples, and Ron ate the rest. What percent of the apples did Ron eat? _____

17. There are 9 sophomores on the basketball team. This is 300% of the number of freshmen on the team. Find the number of freshmen on the team. _____

18. The Eastside Trail is 20 miles long. This is 140% as long as the Westside Trail. Find the length of the Westside Trail to the nearest mile. _____

Name _____ Date _____ Class _____

LESSON 6-3 Practice B
Finding Percents

Find each percent.

1. What percent of 84 is 21?

2. 24 is what percent of 60?

3. What percent of 150 is 75?

4. What percent of 80 is 68?

5. 36 is what percent of 80?

6. What percent of 88 is 33?

7. 19 is what percent of 95?

8. 28.8 is what percent of 120?

9. What percent of 56 is 49?

10. What percent of 102 is 17?

11. What percent of 94 is 42.3?

12. 90 is what percent of 75?

13. Daphne bought a used car for $9200. She made a down payment of $1840. Find the percent of the purchase price that is the down payment.

14. Tricia read $\frac{1}{4}$ of her book on Monday. On Tuesday, she read 36% of the book. On Wednesday, she read 0.27 of the book. She finished the book on Thursday. What percent of the book did she read on Thursday?

15. An airplane traveled from Boston to Las Vegas making a stop in St. Louis. The plane traveled 2410 miles altogether, which is 230% of the distance from Boston to St. Louis. Find the distance from Boston to St. Louis to the nearest mile.

16. The first social studies test had 16 questions. The second test had 220% as many questions as the first test. Find the number of questions on the second test.

Holt Mathematics

Name _____ Date _____ Class _____

LESSON 6-3
Practice C
Finding Percents

Find each number.

1. What number is 80% of 200?

   _____

2. What number is $33\frac{1}{3}$% of 243?

   _____

3. What number is 3.5% of 240?

   _____

4. What number is 265% of 80?

   _____

Complete each statement.

5. Since 12 is 15% of 80,

   24 is _____% of 80.

   36 is _____% of 80.

   48 is _____% of 80.

6. Since 15 is 12% of 125,

   30 is _____% of 125.

   45 is _____% of 125.

   60 is _____% of 125.

7. Since 81 is 150% of 54,

   81 is _____% of 81.

   81 is _____% of 108.

   81 is _____% of 324.

8. Since 18 is 5% of 360,

   18 is _____% of 180.

   18 is _____% of 90.

   18 is _____% of 45.

9. Jerry collected 130 souvenirs from 40 states in the United States and 120 from 20 other countries. What percent of the Jerry's souvenirs are from the United States? _____

10. Dan completed $\frac{3}{8}$ of the science poster. Becky completed 24.5% of the poster, and Greg completed 0.17 of the poster. Amber completed the rest of the poster. What percent of the poster did Amber complete? _____

11. The longest river in the United States is the Missouri River. It is 2,540 miles long, which is 134% of the length of the Rio Grande River. Find the length of the Rio Grande to the nearest ten feet. _____

Name _____ Date _____ Class _____

LESSON 6-3 Reteach
Finding Percents

Since a percent is a ratio, problems involving percent can be solved by using a proportion.

$$\frac{\text{symbol number}}{100} = \frac{\text{is number}}{\text{of number}}$$

There are different possibilities for an unknown quantity in this proportion.

Possibility 1: Find the *symbol number*.

What percent of 80 is 16?

$$\frac{\text{symbol number}}{100} = \frac{\text{is number}}{\text{of number}}$$

$$\frac{x}{100} = \frac{16}{80}$$

$$80 \cdot x = 16 \cdot 100$$

$$\frac{80x}{80} = \frac{1600}{80}$$

$$x = 20 \quad \text{So, 16 is 20\% of 80.}$$

Find what percent one number is of another.

1. What percent of 64 is 16?

    $\frac{x}{100} =$ _____

    _____ • $x =$ _____ • _____

    $\frac{x}{\rule{1cm}{0.4pt}} =$ _____

    $x =$ _____

    So, 16 is _____ of 64.

2. What percent of 200 is 150?

    $\frac{x}{\rule{1cm}{0.4pt}} =$ _____

    _____ • $x =$ _____ • _____

    $\frac{x}{\rule{1cm}{0.4pt}} =$ _____

    $x =$ _____

    So, 150 is _____ of 200.

3. What percent of 4 is 6?

    $\frac{x}{\rule{1cm}{0.4pt}} =$ _____

    _____ • $x =$ _____ • _____

    $\frac{x}{\rule{1cm}{0.4pt}} =$ _____

    $x =$ _____

    So, 6 is _____ of 4.

4. About what percent of 115 is 40?

    $\frac{\rule{1cm}{0.4pt}}{100} =$ _____

    _____ • $x =$ _____ • _____

    $\frac{x}{\rule{1cm}{0.4pt}} =$ _____

    $x =$ _____

    So, 40 is about _____ of 115.

LESSON 6-3

Reteach
Finding Percents (continued)

Possibility 2: Find the *is* number.

What is 20% of 80?

$$\frac{\text{symbol number}}{100} = \frac{\textit{is}\text{ number}}{\textit{of}\text{ number}}$$

$$\frac{20}{100} = \frac{x}{80}$$

$$100 \cdot x = 20 \cdot 80$$

$$\frac{100x}{100} = \frac{1600}{100}$$

$$x = 16 \qquad \text{So, 20\% of 80 is 16.}$$

Find the indicated percent of each number.

5. What is 30% of 150?

$$\frac{\underline{}}{100} = \frac{x}{\underline{}}$$

$$\underline{} \cdot x = \underline{} \cdot \underline{}$$

$$\frac{\underline{} x}{\underline{}} = \frac{\underline{}}{\underline{}}$$

$$x = \underline{}$$

So, 30% of 150 is _____.

6. What is 75% of 205?

$$\frac{\underline{}}{100} = \frac{x}{\underline{}}$$

$$\underline{} \cdot x = \underline{} \cdot \underline{}$$

$$\frac{\underline{} x}{\underline{}} = \frac{\underline{}}{\underline{}}$$

$$x = \underline{}$$

So, 75% of 205 is _____.

7. What is 125% of 300?

$$\frac{\underline{}}{100} = \frac{x}{\underline{}}$$

$$\underline{} \cdot x = \underline{} \cdot \underline{}$$

$$\frac{\underline{} x}{\underline{}} = \frac{\underline{}}{\underline{}}$$

$$x = \underline{}$$

So, 125% of 300 is _____.

8. What is $66\frac{2}{3}$% of 81?

$$\frac{66\frac{2}{3}}{100} = \frac{x}{\underline{}}$$

$$\underline{} \cdot x = 66\frac{2}{3} \cdot \underline{}$$

$$\underline{} \cdot x = \frac{200}{3} \cdot \underline{}$$

$$\underline{} \cdot x = \underline{}$$

$$\frac{\underline{} x}{\underline{}} = \frac{\underline{}}{\underline{}}$$

$$x = \underline{}$$

Name _____ Date _____ Class _____

Challenge
LESSON 6-3 *Particular Percentages*

The shaded portion of this rectangle represents $\frac{1}{4}$, or 25%, of the whole rectangle.

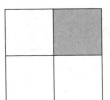

**Find the percent of each figure that is shaded.
Lines intersecting sides or circumferences divide them equally.**

1.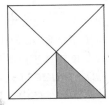

2.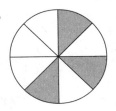

3.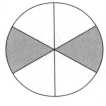

4.

5.

6.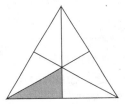

7.

8.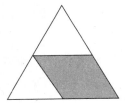

9.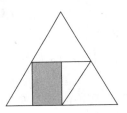

10.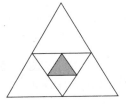

11.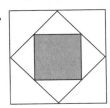

12.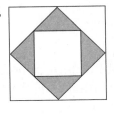

Holt Mathematics

Name _____ Date _____ Class _____

LESSON 6-3 Problem Solving
Finding Percents

Write the correct answer.

1. Florida State University in Tallahassee, Florida has 29,820 students. Approximately 60% of the students are women. How many of the students are women?

2. The yearly cost of tuition, room and board at Florida State University for a Florida resident is $10,064. If tuition is $3,208 a year, what percent of the yearly cost is tuition? Round to the nearest tenth of a percent.

3. The yearly cost of tuition, room and board at Florida State University for a non-Florida resident is $23,196. If tuition is $16,340 a year, what percent of the yearly cost is tuition for a non-resident? Round to the nearest tenth of a percent.

4. Approximately 65% of the students who apply to Florida State University are accepted. If 15,000 students apply to Florida State University, how many would you expect to be accepted?

The top four NBA field goal shooters for the 2003–2004 regular season are given in the table below. Choose the letter for the best answer.

5. What percent of field goals did Shaquille O'Neal make? Round to the nearest tenth of a percent.
   A 0.6%   C 58.4%
   B 1.71%   D 59.2%

6. How many field goals did Donyell Marshall make in the 2003–2004 regular season?
   F 38   H 295
   G 114   J 342

**NBA Field Goal Leaders
2003–2004 Season**

Player	Attempts	Made	Percent
Shaquille O'Neal	948	554	
Donyell Marshall	604		56.6
Elton Brand	950	348	
Dale Davis	913		53.5

7. What percent of field goals did Elton Brand make? Round to the nearest tenth of a percent.
   A 1.87%   C 51.9%
   B 50%   D 36.6%

8. How many field goals did Dale Davis make in the 2003–2004 regular season?
   F 274   H 457
   G 378   J 488

Name _____ Date _____ Class _____

Reading Strategies
LESSON 6-3 *Read a Diagram*

You can use this information to help you solve percent problems.
- When something is unknown, it is shown as a letter.
- The word *of* means to multiply.
- The word *is* means "is equal to."

The diagrams below show you how to set up two different types of percent problems.

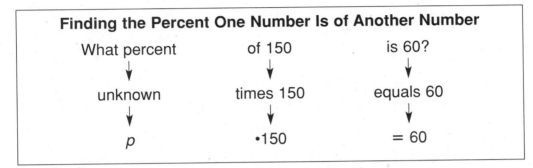

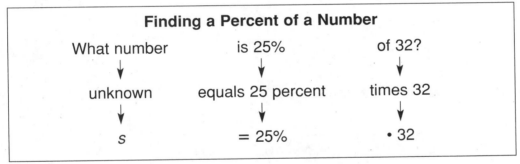

Answer each question.

1. What word means "to multiply"?

   _____

2. What symbol is used to stand for a piece of information that is not known?

   _____

Rewrite the percent problems using symbols.

3. What percent of 70 is 35?

   _____

4. 12 is what percent of 120?

   _____

Puzzles, Twisters & Teasers
LESSON 6-3 Dino-math!

Fill in the blanks to complete each statement. Then use your answers to match the letter to the riddle.

1. 18 is _____ % of 60   Y
2. 8 is _____ % of 80   W
3. 20 is _____ % of 40   D
4. 12.5 is _____ % of 1250 A
5. 150 is _____ % of 80   N
6. 35 is _____ % of 71   S
7. 6 is _____ % of 44   V
8. 25 is _____ % of 150   N
9. 74 is _____ % of 222   I
10. 225 is _____ % of 1130 T

Why did the dinosaur cross the road?

Because the chicken __ __ __ __ , __ N __ E __ T E __ __ E T
                    10 1 49.3 16.7 19.9 33.3 13.6 187.5   50 30

Name _____ Date _____ Class _____

LESSON 6-4 Practice A
Finding a Number When the Percent Is Known

Write an equation or proportion that you can use to find each number.

1. 9 is 15% of what number?

2. 80% of what number is 20?

Find each number.

3. 16 is 64% of what number?

4. 50% of what number is 12?

5. 3 is 5% of what number?

6. 7 is 35% of what number?

7. 75% of what number is 15?

8. 8 is 25% of what number?

9. 60% of what number is 18?

10. 16 is 40% of what number?

11. 7 is 17.5% of what number?

12. 25% of what number is 9.25?

13. 24 is 120% of what number?

14. $33\frac{1}{3}$% of what number is 13?

15. At football practice Ty made 24 field goals, which was 80% of the field goals he attempted. How many field goals did he attempt?

16. At Washington Heights School there are 84 eighth graders, which is 8% of the school's enrollment. How many students attend Washington Heights School?

17. Jackie spends 25% of her monthly salary on rent. Her rent is $650 a month. What is Jackie's monthly salary?

18. Lisa's starting pay was $8.50 per hour. After 8 months she was given a 6% increase. How much per hour was Lisa's raise?

Holt Mathematics

Name _____ Date _____ Class _____

LESSON 6-4 Practice B
Finding a Number When the Percent Is Known

Find each number to the nearest tenth.

1. 40% of what number is 18?
   _____

2. 28 is 35% of what number?
   _____

3. 21 is 60% of what number?
   _____

4. 25% of what number is 19?
   _____

5. 40% of what number is 22?
   _____

6. 41 is 50% of what number?
   _____

7. 50 is 15% of what number?
   _____

8. 0.3% of what number is 24?
   _____

9. 36 is 30% of what number?
   _____

10. 26 is 75% of what number?
    _____

11. 12.5% of what number is 14?
    _____

12. 25% of what number is 28.25?
    _____

13. 27 is $33\frac{1}{3}$% of what number?
    _____

14. 54 is 150% of what number?
    _____

15. There were 546 students at a school assembly. This was 65% of all students who attend Content Middle School. How many students attend Content Middle School?

    _____

16. On his last test Greg answered 64 questions correctly. This was 80% of the questions. How many questions were on the test?

    _____

17. The price of a jacket is $48. If the sales tax rate is 5.5%, what is the amount of tax? What is the total cost of the jacket?

    _____

18. Carla has finished swimming 14 laps in swim practice. This is 70% of the total number of laps she must swim. How many more laps must Carla swim to complete her practice?

    _____

Practice C
Lesson 6-4: Finding a Number When the Percent Is Known

Find each number to the nearest tenth.

1. 65% of what number is 78? _____

2. 112 is 75% of what number? _____

3. 175% of what number is 70? _____

4. 46 is 2% of what number? _____

5. 8 is 125% of what number? _____

6. 0.25% of what number is 54? _____

Complete each statement

7. Since 3% of 400 is 12,

    6% of _____ is 12.

    12% of _____ is 12.

    24% of _____ is 12.

8. Since 100% of 32 is 32,

    80% of _____ is 32.

    50% of _____ is 32.

    25% of _____ is 32.

9. Since 15% of 600 is 90,

    30% of _____ is 90.

    60% of _____ is 90.

    120% of _____ is 90.

10. Since 40% of 60 is 24,

    30% of _____ is 24.

    20% of _____ is 24.

    10% of _____ is 24.

11. Parker bought a suit on sale for $162. He paid 72% of the regular price. What was the regular price of the suit?

    _____

12. Ricardo and Tyler went out for dinner. Tyler's dinner cost $18.50 and Ricardo's dinner cost $17.50. They must pay 6% tax on the meals. The two men also want to leave a 20% tip. They decide to divide the bill evenly. How much will each pay? (*Hint*: do not pay tax on the tip.)

    _____

Name _____ Date _____ Class _____

LESSON 6-4 Reteach
Finding a Number When the Percent Is Known

Since a percent is a ratio, problems involving percent can be solved by using a proportion.

$$\frac{\text{symbol number}}{100} = \frac{\text{is number}}{\text{of number}}$$

To find a number when the percent is known, the variable appears in the *of* position in the proportion

16 is 20% of what number?

$$\frac{20}{100} = \frac{16}{x}$$

$$20 \cdot x = 16 \cdot 100$$

$$\frac{20x}{20} = \frac{1600}{20}$$

$$x = 80$$

So, 16 is 20% of 80.

Find each number whose percentage is given.

1. 18 is 75% of what number?

   $$\frac{}{100} = \frac{}{x}$$

   _____ • x = _____ • _____

   $$\frac{x}{} = \frac{}{}$$

   x = _____

   So, 18 is 75% of _____.

2. 96 is 40% of what number?

   $$\frac{}{100} = \frac{}{x}$$

   _____ • x = _____ • _____

   $$\frac{x}{} = \frac{}{}$$

   x = _____

   So, 96 is 40% of _____.

3. 7 is 125% of what number

   $$\frac{}{100} = \frac{7}{x}$$

   _____ • x = _____ • _____

   $$\frac{x}{} = \frac{}{}$$

   x = _____

   So, 7 is 125% of _____.

4. 40 is about 30% of what number?

   $$\frac{}{100} = \frac{}{}$$

   _____ • x = _____ • _____

   $$\frac{x}{} = \frac{}{}$$

   x = _____

   So, 40 is about 30% of _____.

Copyright © by Holt, Rinehart and Winston.
All rights reserved.

Holt Mathematics

Challenge
Lesson 6-4: In the Chemistry Laboratory

When a chemist dilutes pure acid with another substance, the resulting mixture is no longer pure acid.

Consistent with the words, *pure acid* is 100% acid. So, there are 20 grams of pure acid in 20 grams of a pure-acid solution.

Laura, a chemist, has 20 grams of a solution that is only 40% acid.

1. How many grams of pure acid are there in Laura's acid solution?

Suppose, now, Laura wants to increase the acid content of the 40% acid solution to make it a 50%-acid solution.

2. What do you think Laura has to do to increase the acid content of the solution?

Laura decides to add n ounces of pure acid to increase the acid content of the original 20 grams of 40%-acid solution to make it a 50%-acid solution.

3. Represent in terms of n the total number of grams in the new solution.

4. Represent in terms of n the number of grams of pure acid in the new solution.

Then, the amount of pure acid in the original solution plus the amount of pure acid added equals the amount of pure acid in the new solution.

5. Use the results of Exercises 1 and 4 to write an equation that will find the number n of grams of pure acid that will be added to the original solution to increase its acid content from 40% to 50%. Solve the equation.

6. Explain how to check your result.

Name _____ Date _____ Class _____

Problem Solving
LESSON 6-4 Finding a Number When the Percent is Known

Write the correct answer.

1. The two longest running Broadway shows are *Cats* and *A Chorus Line*. *A Chorus Line* had 6137, or about 82% of the number of performances that *Cats* had. How many performances of *Cats* were there?

2. *Titanic* and *Star Wars* have made the most money at the box office. *Star Wars* made about 76.7% of the money that *Titanic* made at the box office. If *Star Wars* made about $461 million, how much did *Titanic* make? Round to the nearest million dollars.

Use the table below. Round to the nearest tenth of a percent.

3. What percent of students are in Pre-K through 8th grade?

4. What percent of students are in grades 9–12?

Public Elementary and Secondary School Enrollment, 2001

Grades	Population (in thousands)
Pre-K through grade 8	33,952
Grades 9–12	13,736
Total	47,688

Choose the letter for the best answer.

5. In 2000, women earned about 72.2% of what men did. If the average woman's weekly earnings was $491 in 2000, what was the average man's weekly earnings? Round to the nearest dollar.
   A $355   C $680
   B $542   D $725

6. The highest elevation in North America is Mt. McKinley at 20,320 ft. The highest elevation in Australia is Mt. Kosciusko, which is about 36% of the height of Mt. McKinley. What is the highest elevation in Australia? Round to the nearest foot.
   F 5480 ft   H 12,825 ft
   G 7315 ft   J 56,444 ft

7. The Gulf of Mexico has an average depth of 4,874 ft. This is about 36.2% of the average depth of the Pacific Ocean. What is the average depth of the Pacific Ocean? Round to the nearest foot.
   A 1764 ft   C 10,280 ft
   B 5843 ft   D 13,464 ft

8. Karl Malone is the NBA lifetime leader in free throws. He attempted 11,703 and made 8,636. What percent did he make? Round to the nearest tenth of a percent.
   F 1.4%   H 73.8%
   G 58.6%   J 135.6%

Holt Mathematics

Name _____ Date _____ Class _____

LESSON 6-4 Reading Strategies
Connecting Words and Symbols

To write equations for percent problems, connect words with symbols. Then change percents to decimals.

28 is 10% of what number? ← words
28 = 10% · n   ← symbols
28 = 0.10 · n   ← decimal form

Answer each question.

1. What is the decimal form of 10%?

    _____

2. What symbol stands for "what number"?

    _____

3. What symbol stands for "of"?

    _____

75 is 50% of what number? ← words
75 = 50% · n   ← symbols
75 = 0.50 · n   ← decimal form

Answer each question.

4. What is the decimal form of 50%?

    _____

5. What symbol stands for "is"?

    _____

6. Write the words that mean n.

    _____

7. Write the symbols for this percent problem: 50 is 6% of what number?

    _____

35

Holt Mathematics

Name _____ Date _____ Class _____

LESSON 6-4 Puzzles, Twisters & Teasers
Your Lucky Number!

Black out all the INCORRECT statements. What you see from the pattern created will be your lucky number.

5 is 30% of 70	57 is 10% of 570	24 is 15% of 160	8 is 25% of 32	6 is 2% of 300
54 is 25% of 300	9 is 10% of 99	45 is 50% of 100	78 is 10 % of 7	4 is 40% of 10
21 is 75% of 210	16 is 49% of 140	968 is 10% of 96	33 is 60% of 500	51 is 6% of 850
35 is 15 % of 70	44 is 10% of 88	76 is 50% of 1050	22 is 40% of 88	77 is 25% of 308
65 is 100% of 6.5	43 is 67% of 536	7 is 15 % of 734	18 is 10 % of 230	97 is 30% of 323
24 is 13% of 96	74 is 30% of 300	640 is 5% of 100	10 is 10% of 1000	30 is 18% of 167
55 is 87 % of 230	3 is 50 % of 30	19 is 43 % of 87	86 is 34 % of 68	20 is 20% of 100

What is your lucky number? _____

Name _____ Date _____ Class _____

Practice A
LESSON 6-5
Percent Increase and Decrease

State whether each change represents an increase or decrease.

1. from 10 to 15
2. from 16 to 12
3. from 8 to 14

_____ _____ _____

Find each percent increase or decrease to the nearest percent.

4. from 2 to 5
5. from 10 to 6
6. from 12 to 18

_____ _____ _____

7. from 8 to 5.6
8. from 15 to 8
9. from 21 to 15

_____ _____ _____

10. from 17 to 21
11. from 10 to 2
12. from 4 to 9

_____ _____ _____

13. from 7 to 11
14. from 3 to 9
15. from 12 to 5

_____ _____ _____

16. World Toys buys bicycles for $38 and sells them for $95. What is the percent of increase in the price? _____

17. Jack bought a stereo on sale for $231. The original price was $385. What was the percent of decrease in price? _____

18. Adams Clothing Store buys coats for $50 and sells them for $80. What percent of increase is this? _____

19. Asabi's average in math for the first quarter of the school year was 75. His second quarter average was 81. What was the percent of increase in Asabi's grade? _____

20. A shoe store is selling athletic shoes at 30% off the regular price. If the regular price of a pair of athletic shoes is $45, what is the sale price? _____

Practice B
6-5 Percent Increase and Decrease

Find each percent increase or decrease to the nearest percent.

1. from 16 to 20
2. from 30 to 24
3. from 15 to 30

4. from 35 to 21
5. from 40 to 46
6. from 45 to 63

7. from 18 to 26.1
8. from 24.5 to 21.56
9. from 90 to 72

10. from 29 to 54
11. from 42 to 92.4
12. from 38 to 33

13. from 64 to 36.4
14. from 78 to 136.5
15. from 89 to 32.9

16. Mr. Havel bought a car for $2400 and sold it for $2700. What was the percent of profit for Mr. Havel in selling the car? _____

17. A computer store buys a computer program for $24 and sells it for $91.20. What is the percent of increase in the price? _____

18. A manufacturing company with 450 employees begins a new product line and must add 81 more employees. What is the percent of increase in the number of employees? _____

19. Richard earns $2700 a month. He received a 3% raise. What is Richard's new annual salary? _____

20. Marlis has 765 cards in her baseball card collection. She sells 153 of the cards. What is the percent of decrease in the number of cards in the collection? _____

Name _____ Date _____ Class _____

LESSON 6-5 Practice C
Percent Increase and Decrease

Find each percent increase or decrease to the nearest percent.

1. from 120 to 162

2. from 84 to 47.04

3. from 72 to 46.8

_____

_____

_____

4. from 90 to 189

5. from 67 to 112

6. from 153 to 109

_____

_____

_____

Find each missing number.

7. originally: $300

    new price: $450

    _____% increase

8. originally: $850

    new price: $1147.50

    _____% increase

9. originally: $2500

    new price: $825

    _____% decrease

10. originally: $_____

    new price: $840

    60% decrease

11. originally: $200

    new price: $_____

    137% increase

12. originally: $4.20

    new price: $6.09

    _____% increase

13. Fandango Store buys a computer program for $244. It sells the computer program for $927.20. What is the percent of increase in the price?

_____

14. Denise buys a shirt on sale for $21.08. This represents a 15% decrease in price. What was the original price of the shirt?

_____

15. A storeowner purchases 40 shirts for $600. She then adds 40% to her cost and tags each shirt with the same selling price. What is the amount of profit for each shirt?

_____

LESSON 6-5 Reteach
Percent Increase and Decrease

To find the percent increase:
- Find the amount of increase by subtracting the lesser number from the greater.
- Write a fraction: **percent increase** = $\dfrac{\text{amount of increase}}{\text{original amount}}$
- If possible, simplify the fraction.
- Rewrite the fraction as a percent.

The temperature increased from 60°F to 75°F.
Find the percent of increase.

percent of increase = $\dfrac{75° - 60°}{60°} = \dfrac{15°}{60°} = \dfrac{1}{4} = 25\%$

Complete to find each percent increase.

1. Membership increased from 80 to 100.

   ___ − ___

   = ___

   $\dfrac{\_\_\_}{80} = \dfrac{\_\_\_}{\_\_\_}$

   = ___ %

2. Savings increased from $500 to $750.

   ___ − ___

   = ___

   $\dfrac{\_\_\_}{500} = \dfrac{\_\_\_}{\_\_\_}$

   = ___ %

   Find the amount of increase.

   percent increase = $\dfrac{\text{amount of increase}}{\text{original amount}}$

   Change the fraction to a percent.

3. Price increased from $20 to $23.

   _____

   $\dfrac{\_\_\_}{20}$

   $\dfrac{\_\_\_}{20} = 20\overline{)\_\_\_\_} = \_\_\_ \%$

   Find the amount of increase.

   percent increase = $\dfrac{\text{amount of increase}}{\text{original amount}}$

   Change the fraction to a percent.

Reteach

LESSON 6-5 Percent Increase and Decrease (continued)

To find the percent decrease:
- Find the amount of decrease by subtracting the lesser number from the greater.
- Write a fraction: **percent decrease** = $\dfrac{\text{amount of decrease}}{\text{original amount}}$
- If possible, simplify the fraction.
- Rewrite the fraction as a percent.

Carl's weight decreased from 175 lb to 150 lb.
Find the percent of decrease.

percent of decrease = $\dfrac{175 - 150}{175} = \dfrac{25}{175} = \dfrac{1}{7} = 7\overline{)1.000}^{\,0.143} = 14.3\%$

Complete to find each percent decrease.

4. Enrollment decreased from 1000 to 950.

   ___ − ___

   = ___

   $\dfrac{}{1000} = \dfrac{}{100}$

   = ___ %

5. Temperature decreased from 75°F to 60°F.

   ___ − ___   Find the amount of decrease.

   = ___

   $\dfrac{}{75} = \dfrac{}{15} = \dfrac{}{100}$ percent decrease = $\dfrac{\text{amount of decrease}}{\text{original amount}}$

   = ___ %   Change the fraction to a percent.

6. Sale price decreased from $22 to $17.

   _____   Find the amount of decrease.

   $\dfrac{}{22}$ percent decrease = $\dfrac{\text{amount of decrease}}{\text{original amount}}$

   $\dfrac{}{22} = 22\overline{)} = $ ___ %   Change the fraction to a percent.

Holt Mathematics

Name _____ Date _____ Class _____

LESSON 6-5 Challenge
The Ups and Downs of the Marketplace

Prices change. The price of a stock can change every few minutes. The price of a house changes over a longer period of time.

The *selling price* of an item is what someone is willing to pay for it. It is a good measure of market value.

Find the current value of each item. Round your answer to the nearest cent

1. a. Amy bought a baseball card for $12. To date, the value of the card increased by 30%, then decreased by 15%, and finally increased by 40%.

Joe bought a baseball card for $12. To date, the value of the card decreased by 10%, then increased by 70%, and finally decreased by 5%.

Whose card is currently worth more? by how much? Explain.

_____
_____
_____

b. By about what percent must the currently lesser-valued card increase to be of equal value with the greater-valued card? Round your answer to the nearest tenth of a percent.

_____

2. a. Jorge's family bought a house for $125,000. To date, the value of the house increased by 5%, then decreased by 25%, and finally increased by 10%.

Gene's family bought a house for $125,000. To date, the value of the house decreased by 5%, then increased by 15%, and finally decreased by 20%.

Whose house is currently worth more? by how much? Explain.

_____
_____
_____

b. By about what percent must the currently lesser-valued house increase to be of equal value with the greater-valued house? Round your answer to the nearest tenth of a percent.

_____

Name _____ Date _____ Class _____

Problem Solving
LESSON 6-5 Percent Increase and Decrease

Use the table below. Write the correct answer.

1. What is the percent increase in the population of Las Vegas, NV from 1990 to 2000? Round to the nearest tenth of a percent.

   _____

2. What is the percent increase in the population of Naples, FL from 1990 to 2000? Round to the nearest tenth of a percent.

   _____

Fastest Growing Metropolitan Areas, 1990–2000

Metropolitan Area	Population 1990	Population 2000	Percent Increase
Las Vegas, NV	852,737	1,563,282	
Naples, FL	152,099	251,377	
Yuma, AZ	106,895		49.7%
McAllen-Edinburg-Mission, TX	383,545		48.5%

3. What was the 2000 population of Yuma, AZ to the nearest whole number?

   _____

4. What was the 2000 population of McAllen-Edinburg-Mission, TX metropolitan area to the nearest whole number?

   _____

For exercises 5–7, round to the nearest tenth. Choose the letter for the best answer.

5. The amount of money spent on automotive advertising in 2000 was 4.4% lower than in 1999. If the 1999 spending was $1812.3 million, what was the 2000 spending?

   A $79.7 million   C $1892 million
   B $1732.6 million D $1923.5 million

6. In 1967, a 30-second Super Bowl commercial cost $42,000. In 2000, a 30-second commercial cost $1,900,000. What was the percent increase in the cost?

   F 1.7%   H 442.4%
   G 44.2%   J 4423.8%

7. In 1896 Thomas Burke of the U.S. won the 100-meter dash at the Summer Olympics with a time of 12.00 seconds. In 2004, Justin Gatlin of the U.S. won with a time of 9.85 seconds. What was the percent decrease in the winning time?

   A 2.15%   C 21.8%
   B 17.9%   D 45.1%

8. In 1928 Elizabeth Robinson won the 100-meter dash with a time of 12.20 seconds. In 2004, Yuliya Nesterenko won with a time that was about 10.4% less than Robinson's winning time. What was Nesterenko's time, rounded to the nearest hundredth?

   F 9.83 seconds   H 12.16 seconds
   G 10.93 seconds   J 13.47 seconds

Name _____ Date _____ Class _____

LESSON 6-5 Reading Strategies
Compare and Contrast

Percent can be used to describe change. It is shown as a ratio.

$$\text{percent of change} = \frac{\text{amount of change}}{\text{original amount}}$$

Compare the two lists. Change can either increase or decrease.

Increase	Decrease
A collector sold 15 CDs. Then she sold 25 more CDs.	Ben had a collection of 60 CDs. Now he has only 45 CDs.
↓	↓
Sales went up, so the ratio will show a **percent of increase**.	The CD collection went down, so the ratio will show a **percent of decrease**.
Change: 25 − 15 = 10 more CDs	Change: 60 − 45 = 15 fewer CDs
Percent of change = $\frac{10}{25}$	Percent of change = $\frac{15}{60}$
Change fraction to percent: 40%	Change fraction to percent: 25%

1. Compare percent of increase with percent of decrease. How are they the same?

   _____

2. Write the ratio that stands for percent of change.

   _____

Write *percent of increase* or *percent of decrease* to describe each situation.

3. Sophie had $70 saved. She withdrew $15 from her savings.

   _____

4. Kate bought $50 worth of groceries. Then she bought $20 more.

   _____

Name _____ Date _____ Class _____

Puzzles, Twisters & Teasers
6-5 Do Chickens Have Funny Bones?

Circle words from the list that you find.

Find a word that answers the riddle. Circle it and write it on the line.

percent   change   increase   decrease   ratio
amount   original decimal   application describe

```
A P P L I C A T I O N D F R
M V N B D E C R E A S E C V
O R I G I N A L A S D C E R
U A Z X C V B N M K O I J I
N T P E R C E N T E W M B Y
T I N C R E A S E U I A L O
C O R N Y Q W E R T Y L C E
C H A N G E D E S C R I B E
```

What kind of jokes do chickens like best?

_____ ones

Name _____ Date _____ Class _____

Practice A
LESSON 6-6 Applications of Percents

Let c = the commision amount and write an equation to find the commission for the following. Do not solve.

1. 10% commission on $4000

2. 6% commission on $8450

3. 8% commission on $3575

4. 12% commission on $12,750

5. 5.5% commission on $60,000

6. $6\frac{1}{4}$% commission on $85,900

Write a proportion to represent the following. Do not solve.

7. What percent of 14 is 7?

8. 7 is what percent of 25?

9. What number is 12.5% of 16?

10. 21 is 35% of what number?

Solve.

11. 45 is 25% of what number?

12. What percent of 288 is 36?

13. A financial investment broker earns 4% on each customer dollar invested. If the broker invests $50,000, what is the commission on the investment? _____

14. Sharlene bought 4 CDs at the music store. Each cost $14.95. She was charged 5% sales tax on her purchase. What was the total cost of her purchase? _____

15. Isaac earned $1,800 last month. He put $270 into savings. What percent of his earnings did Isaac put in savings? _____

16. Edel works for a company that pays a 15% commission on her total sales. If she wants to earn $450 in commissions, how much do her total sales have to be? _____

Name _____ Date _____ Class _____

LESSON 6-6 Practice B
Applications of Percents

Complete the table to find the amount of sales tax for each sale amount to the nearest cent.

1.

Sale amount	5% sales tax	8% sales tax	6.5% sales tax
$67.50			
$98.75			
$399.79			
$1250.00			

Complete the table to find the commission for each sale amount to the nearest cent.

2.

Sale amount	6% commision	9% commision	8.5% commission
$475.00			
$2450.00			
$12,500.00			
$98,900.00			

3. Alice earns a monthly salary of $315 plus a commission on her total sales. Last month her total sales were $9640, and she earned a total of $1182.60. What is her commission rate? _____

4. Phillipe works for a computer store that pays a 12% commission and no salary. What will Phillipe's weekly sales have to be for him to earn $360? _____

5. The purchase price of a book is $35.85. The sales tax rate is 6.5%. How much is the sales tax to the nearest cent? What is the total cost of the book?

_____

6. Who made more commission this month? How much did she make? Salesperson A made 11% of $67,530. Salesperson B made 8% of $85,740.

_____

7. Jon earned $38,000 last year. He paid $6,840 towards entertainment. What percent of his earnings did Jon pay in entertainment expenses? _____

8. The Cougars won 62% of their games. They won 93 games. How many games did they lose? _____

Practice C
6-6 Applications of Percents

Find each commission or sales tax, to the nearest cent.

1. total sales $9450
   commission rate: 8%

   _____

2. total sales $21,097
   sales tax rate: 5.5%

   _____

3. total sales $1089
   sales tax rate: $6\frac{1}{8}\%$

   _____

4. total sales $16,772
   commission rate: 15%

   _____

Find the total sales, to the nearest cent.

5. commission: $41.50
   commission rate: 8%

   _____

6. commission: $263.70
   commission rate: $4\frac{1}{2}\%$

   _____

7. commission: $614.25
   commission rate: 6.25%

   _____

8. commission: $2250
   commission rate: 15%

   _____

9. A model car has a list price of $46.20. The model is on sale at 15% off. Find the total cost to the nearest cent after a 4.5% sales tax is added to the sale price.

10. An item priced at $776 has a sales tax of $48.50. Find the sales tax rate expressed as a percent.

11. James earned $39,600 last year. He paid $11,250 towards rent and $1,422 in car payments. What percent of his earnings did Jon pay in rent and car payments?

12. Find the total monthly pay if total sales for a month were $35,450, the commission rate is 4.5%, and the weekly base salary is $175.

13. Ms. Simms is paid an 8% commission on all sales. She had sales of $89,400 for the month. Ms. Harris works for a different company, and also sold $89,400 for the month but made $447 more than Ms. Simms. What is Ms. Harris' commission rate?

Holt Mathematics

Name _____ Date _____ Class _____

Reteach
LESSON 6-6 Applications of Percents

Salespeople often earn a **commission,** a percent of their total sales.

Find the commission on a real-estate sale of $125,000
if the commission rate is 4%.

Write the percent as a decimal and multiply.

    commission rate × amount of sale = amount of commission
          0.04 × $125,000   = $5000

If, in addition to the commission, the salesperson earns a
salary of $1000, what is the total pay?

    commission + salary = total pay
       $5000 + $1000 = $6000

Complete to find each total monthly pay.

1. total monthly sales = $170,000; commission rate = 3%; salary = $1500

   amount of commission = 0.03 × $_____ = $_____

   total pay = $_____ + $1500 = $_____

2. total monthly sales = $16,000; commission rate = 5.5%; salary = $1750

   amount of commission = _____ × $_____ = $_____

   total pay = $_____ + $_____ = $_____

A **tax** is a charge, usually a percentage, generally imposed by a government.

Sales tax is the tax on the sale of an item or service.

If the sales tax rate is 7%, find the tax on a sale of $9.49.

Write the tax rate as a decimal and multiply.

    tax rate × amount of sale = amount of tax
      0.07   × $9.49   = $0.6643 ≈ $0.66

Complete to find each amount of sales tax.

3. item price = $5.19; sales tax rate = 6%

   amount of sales tax = 0.06 × $_____ = $_____ ≈ $_____

4. item price = $250; sales tax rate = 6.75%

   amount of sales tax = _____ × $_____ = $_____ ≈ $_____

LESSON 6-6 Reteach
Applications of Percents (continued)

Use a proportion to find what percent of a person's income goes to a specific expense.

Heather earned $3,200 last month. She paid $448 for transportation. To find the percent of her earnings that she put towards transportation, write a proportion.

Think: What percent of 3200 is 448?

$\dfrac{n}{100} = \dfrac{448}{3200}$ ← Set up a proportion.

Think: $\dfrac{\text{part}}{\text{whole}} = \dfrac{\text{part}}{\text{whole}}$

$3200n = 448 \times 100$ ← Find cross products.

$3200n = 44{,}800$ ← Simplify.

$\dfrac{3200n}{3200} = \dfrac{44{,}800}{3200}$ ← Divide both sides by 3200.

$n = 14$ ← Simplify.

Heather put 14% of her earnings towards transportation.

Complete each proportion to find the percent of earnings.

5. Wayne earned $3,100 last month. He paid $837 for food. What percent of his earnings went to food?

   $\dfrac{n}{100} = \dfrac{\_\_\_\_}{3100}$

   $3100n = \_\_\_\_ \times 100$

   $3100n = \_\_\_\_$

   $\dfrac{3100n}{3100} = \_\_\_\_$

   $n = \_\_\_\_$

   ____ of Wayne's earnings went to food.

6. Leah earned $1,900 last month. She paid $304 for utilities. What percent of her earnings went to utilities?

   $\dfrac{n}{100} = \dfrac{304}{\_\_\_\_}$

   $\_\_\_\_ \times n = \_\_\_\_ \times 100$

   $\_\_\_\_ = \_\_\_\_$

   $\_\_\_\_ = \_\_\_\_$

   $n = \_\_\_\_$

   ____ of Leah's earnings went to utilities.

Name _____ Date _____ Class _____

Challenge
LESSON 6-6 *Shoppers' Delight*

Shoppers save money by buying items on sale.
The amount by which the regular price is reduced is called a **discount**.

amount of discount = discount rate × regular price
sale price = regular price − amount of discount

Find the sale price after each discount.

1. regular price = $899;
   discount rate = 20%

   amount of discount = _____

   sale price = _____

2. regular price = $14.99;
   discount rate = 15%

   amount of discount = _____

   sale price = _____

Stores may offer discounts in a variety of ways.

Use $100 as the regular price for the item to write your explanations.

3. Buy one at regular price. Get a second one for half price. Explain how this is different from getting a 50% discount.

   _____

   _____

4. Buy two. Get one free. Explain how this is different from getting a $33\frac{1}{3}\%$ discount.

   _____

   _____

5. This item is marked down by 10%. Use a coupon and get an additional 10% off. Explain how this is different from getting a 20% discount.

   _____

   _____

   _____

6. An item is marked "50% off — Today Only Get Another 50% off". Explain why the item is not free.

   _____

   _____

Name _____ Date _____ Class _____

LESSON 6-6 Problem Solving
Applications of Percents

Write the correct answer.

1. The sales tax rate for a community is 6.75%. If you purchase an item for $500, how much will you pay in sales tax?

2. A community is considering increasing the sales tax rate 0.5% to fund a new sports arena. If the tax rate is raised, how much more will you pay in sales tax on $500?

3. Trent earned $28,500 last year. He paid $8,265 for rent. What percent of his earnings did Trent pay for rent?

4. Julie has been offered two jobs. The first pays $400 per week. The second job pays $175 per week plus 15% commission on her sales. How much will she have to sell in order for the second job to pay as much as the first?

Choose the letter for the best answer. Round to the nearest cent.

5. Clay earned $2,600 last month. He paid $234 for entertainment. What percent of his earnings did Clay pay in entertainment expenses?
   A 9%
   B 11%
   C 30%
   D 90%

6. Susan's parents have offered to help her pay for a new computer. They will pay 30% and Susan will pay 70% of the cost of a new computer. Susan has saved $550 for a new computer. With her parents help, how expensive of a computer can she afford?
   F $165.00   H $1650.00
   G $785.71   J $1833.33

7. Kellen's bill at a restaurant before tax and tip is $22.00. If tax is 5.25% and he wants to leave 15% of the bill including the tax for a tip, how much will he spend in total?
   A $22.17   C $26.63
   B $26.46   D $27.82

8. The 8th grade class is trying to raise money for a field trip. They need to raise $600 and the fundraiser they have chosen will give them 20% of the amount that they sell. How much do they need to sell to raise the money for the field trip?
   F $120.00   H $3000.00
   G $857.14   J $3200.00

Reading Strategies
6-6 Focus On Vocabulary

A **commission** is a percent of money a person is paid for making a sale. Many salespeople receive a commission on the amount they sell.

The **commission rate** is the percent paid on a sale. A salesperson might receive a 5% commission in addition to his salary. The commission rate is 5%.

The formula for finding out how much a salesperson earns based on the commission rate and the amount of sales is:

commission rate • sales = amount of commission

Sales tax is added to the price of an item or service. Sales tax is a percent of the purchase price. A sales tax of 6.5% means that all taxable items will have an additional 6.5% added to the total cost.

sales tax rate • sale price = sales tax

sale price + sales tax = total sale

The **total sale** price is computed by adding the sales tax to the cost of all the items purchased.

Write *commission*, *commission rate*, *sales tax*, or *total sale* to describe each situation.

1. $5.45 was added to the price of the shoes Jill bought.

2. The man who sold your family a car receives $500 for the sale.

3. Mr. Adams makes a 4% commission on each house he sells.

4. Caroline spent $37.43 for two shirts plus tax.

Name _____ Date _____ Class _____

Puzzles, Twisters & Teasers
LESSON 6-6 *One Cool Cat!*

Circle words from the list that you find.
Find a word that answers the riddle. Circle it and write it on the line.

commission   sales   tax   earnings   rate
equation   percent   decimal   convert   multiply

```
S O U R P U S S M V C D P
A R F U Y H B C U W E E L
L M J I P L U O L O R C M
E Q U A T I O N T A X I N
S S D E T Y I V I A Z M J
R A T E I O P E P N H A I
P E R C E N T R L P O L U
W E R T Y U I T Y S W E B
C O M M I S S I O N V Y G
Z W I T H H O L D I N G Y
Z X C E T I J O P L Y R E
Q E A R N I N G S V N O C
```

What do you call a cat that drinks lemonade?

A _____

Name _____ Date _____ Class _____

LESSON 6-7 Practice A
Simple Interest

Write the formula to compute the missing value. Do not solve.

1. principal = $100
   rate = 4%
   time = 2 years
   interest = ?

2. principal = $150
   rate = ?
   time = 2 years
   interest = $9

3. principal = $200
   rate = 5%
   time = ?
   interest = $10

4. principal = ?
   rate = 3%
   time = 4 years
   interest = 30

5. Jules borrowed $500 for 3 years at a simple interest rate of 6%. How much interest will be due at the end of 3 years? How much will Jules have to repay?

6. Karin maintained a balance of $250 in her savings account for 8 years. The financial institution paid simple interest of 4%. What was the amount of interest earned?

Complete the table.

	Principal	Rate	Time	Interest
7.	$300	3%	4 years	
8.	$450		3 years	$67.50
9.	$500	4.5%		$112.50
10.		8%	2 years	$108
11.	$700	4%	3 years	
12.	$750		2 years	$90
13.	$800	2.5%		$100

Name _____ Date _____ Class _____

LESSON 6-7
Practice B
Simple Interest

Find the missing value.

1. principal = $125
   rate = 4%
   time = 2 years
   interest = ?

2. principal = ?
   rate = 5%
   time = 4 years
   interest = $90

3. principal = $150
   rate = 6%
   time = ? years
   interest = $54

4. principal = $200
   rate = ?%
   time = 3 years
   interest = $30

5. principal = $550
   rate = ?%
   time = 3 years
   interest = $57.75

6. principal = ?
   rate = $3\frac{1}{4}$%
   time = 2 years
   interest = $63.05

7. Kwang deposits money in an account that earns 5% simple interest. He earned $546 in interest 2 years later. How much did he deposit?

8. Simon opened a certificate of deposit with the money from his bonus check. The bank offered 4.5% interest for 3 years of deposit. Simon calculated that he would earn $87.75 interest in that time. How much did Simon deposit to open the account?

9. Douglas borrowed $1000 from Patricia. He agreed to repay her $1150 after 3 years. What was the interest rate of the loan?

10. What is the interest paid for a loan of $800 at 5% annual interest for 9 months?

Name _____ Date _____ Class _____

Practice C
LESSON 6-7 *Simple Interest*

Find the interest and the total amount to the nearest cent.

1. $345 at 4% per year for 3 years

2. $782 at 3.5% per year for 4 years

3. $6125 at 7% per year for 2.5 years

4. $9875 at $3\frac{1}{4}$% per year for 5 years

5. $2065 at 5.5% per year for 42 months

6. $1750 at $6\frac{1}{8}$% per year for 33 months

7. $900 at 11% per year for 3 months

8. $8417 at 18% per year for 1 month

9. Will deposited $1550 in an account that pays $8\frac{3}{4}$% annually. How much would be in the account at the end of 24 months?

10. What is the annual interest rate if $7200 is invested for 15 months and earns $855 interest?

11. How long will it take a deposit of $4500 at an annual rate 5.75% to earn $1035?

12. Noah bought a new car costing $25,350. He made a 20% down payment on the car and financed the remaining cost of the car for 5 years at 6.5%. How much interest did Noah pay on his car loan?

13. Mr. Silva earned $196.50 in interest in a year for an account that paid 3% interest per year. If he did not take any money out of the account during the year, how much was in the account at the start of the year?

Holt Mathematics

Name _____ Date _____ Class _____

LESSON 6-7 Reteach
Simple Interest

Interest is money paid on an investment.
A borrower pays the interest. An investor earns the interest.

Simple interest, I, is earned when
an amount of money, the *principal* P,
is borrowed or invested at a *rate of interest* r
for a *period of time* t.

> **Interest = Principal · Rate · Time**
> $I = P \cdot r \cdot t$

Situation 1: Find I given P, r, and t.

Calculate the simple interest on a loan of $3500
for a period of 6 months at a yearly rate of 5%.

Write the interest rate as a decimal.	5% = 0.05
Write the time period in terms of years.	6 months = 0.5 year

$I = P \cdot r \cdot t$
$I = 3500 \cdot 0.05 \cdot 0.5 = \$87.50$ ← interest earned

Find the interest in each case.

1. principal $P = \$5000$; time $t = 2$ years; interest rate $r = 6\%$

$I = P \cdot r \cdot t =$ _____ $\cdot\ 0.06\ \cdot$ _____ $= \$$ _____

2. principal $P = \$2500$; time $t = 3$ months; interest rate $r = 8\%$

$I = P \cdot r \cdot t =$ _____ $\cdot$ _____ $\cdot$ _____ $= \$$ _____

Situation 2: Find t given I, P, and r.

An investment of $3000 at a yearly rate
of 6.5% earned $390 in interest. Find
the period of time for which the money
was invested.

$I = P \cdot r \cdot t$
$390 = 3000 \cdot 0.065 \cdot t$
$390 = 195t$
$\dfrac{390}{195} = \dfrac{195t}{195}$
$2 = t$

The investment was for 2 years.

Find the time in each case.

3. $I = \$1120$; $P = \$4000$; $r = 7\%$
$I = P \cdot r \cdot t$
$1120 =$ _____ $\cdot\ 0.07\ \cdot t$
$1120 =$ _____ t
$\dfrac{\_\_\_}{\_\_\_} = \dfrac{\_\_\_ t}{\_\_\_}$
_____ years $= t$

4. $I = \$812.50$; $P = \$5000$; $r = 6.5\%$
$I = P \cdot r \cdot t$
$812.50 =$ _____ $\cdot$ _____ $\cdot t$
$812.50 =$ _____ t
$\dfrac{\_\_\_}{\_\_\_} = \dfrac{\_\_\_ t}{\_\_\_}$
_____ years $= t$

Name _____ Date _____ Class _____

LESSON 6-7 Reteach
Simple Interest

Situation 3: Find r given I, P, and t.
$2500 was invested for 3 years and earned $450 in interest. Find the rate of interest.

$I = P \cdot r \cdot t$
$450 = 2500 \cdot r \cdot 3$
$450 = 7500r$
$\dfrac{450}{7500} = \dfrac{7500r}{7500}$
$0.06 = r$

The interest rate was 6%.

Find the interest rate in each case.

5. I = $1200; P = $6000; t = 4 years
$I = P \cdot r \cdot t$
$1200 = $ _____ $\cdot r \cdot 4$
$1200 = $ _____ r

$\dfrac{\rule{1cm}{0.4pt}}{\rule{1cm}{0.4pt}} = \dfrac{\rule{0.5cm}{0.4pt} r}{\rule{1cm}{0.4pt}}$

_____ = r

The interest rate was _____ %.

6. I = $325; P = $2000; t = 2.5 years
$I = P \cdot r \cdot t$
$325 = $ _____ $\cdot r \cdot$ _____
$325 = $ _____ r

$\dfrac{\rule{1cm}{0.4pt}}{\rule{1cm}{0.4pt}} = \dfrac{\rule{0.5cm}{0.4pt} r}{\rule{1cm}{0.4pt}}$

_____ = r

The interest rate was _____ %.

The total amount A of money in an account after interest has been earned, is the sum of the principal P and the interest I.

Amount = Principal + Interest
$A = P + I$

Find the amount of money in the account after $3500 has been invested for 3 years at a yearly rate of 6%.

First, find the interest earned.
$I = P \cdot r \cdot t$
$I = 3500 \cdot 0.06 \cdot 3 = \$630$ ⟵ interest earned

Then, add the interest to the principal. $3500 + 630 = 4130$
So, the total amount in the account after 3 years is $4130.

Find the total amount in the account.

7. principal P = $4500; time t = 2.5 years; interest rate r = 5.5%

$I = P \cdot r \cdot t = $ _____ $\cdot$ _____ $\cdot$ _____ = $ _____

Total Amount $= P + I = 4500 + $ _____ = _____

So, after 2.5 years, the total amount in the account was $ _____ .

LESSON 6-7 Challenge
Feather Your Nest

In these exercises, you will solve an investment problem algebraically.

Problem Nancy invested a sum of money at 6%.
She invested a second sum, $500 more than the first, at 8%.
The total interest earned for the year was $180.
How much did Nancy invest at each rate?

1. Let x represent the sum Nancy invested at 6%. Write an expression in terms of x for the interest she earned after 1 year from the 6%-investment.

   _____

2. x represents the sum Nancy invested at 6%.

   a. Write an expression in terms of x for the sum she invested at 8%.

   _____

   b. Write an expression in terms of x for the interest she earned after 1 year from the 8%-investment.

   _____

3. Using your results from Exercises 1 and 2b, write an equation in terms of x to show that the total of the interest earned from the two investments is equal to $180.

   _____

4. Follow these steps to solve your equation for x.

   a. Apply the Distributive Property. _____

   b. Collect like terms on the left side. _____

   c. Subtract. _____

   d. Divide to find x. _____

   So, the sum invested at 6% is $ _____

   and the sum invested at 8% is $ _____

5. Explain how to check your result.

   _____

LESSON 6-7: Problem Solving — Simple Interest

Write the correct answer.

1. Joanna's parents agree to loan her the money for a car. They will loan her $5,000 for 5 years at 5% simple interest. How much will Joanna pay in interest to her parents?

2. How much money will Joanna have spent in total on her car with the loan described in exercise 1?

3. A bank offers simple interest on a certificate of deposit. Jaime invests $500 and after one year earns $40 in interest. What was the interest rate on the certificate of deposit?

4. How long will Howard have to leave $5000 in the bank to earn $250 in simple interest at 2%?

Jan and Stewart Jones plan to borrow $20,000 for a new car. They are trying to decide whether to take out a 4-year or 5-year simple interest loan. The 4-year loan has an interest rate of 6% and the 5-year loan has an interest rate of 6.25%. Choose the letter for the best answer.

5. How much will they pay in interest on the 4-year loan?
   A $4500
   B $4800
   C $5000
   D $5200

6. How much will they repay with the 4-year loan?
   F $24,500
   G $24,800
   H $25,000
   J $25,200

7. How much will they pay in interest on the 5-year loan?
   A $5000
   B $6000
   C $6250
   D $6500

8. How much will they repay with the 5-year loan?
   F $25,000
   G $26,000
   H $26,250
   J $26,500

9. How much more interest will they pay with the 5-year loan?
   A $1000
   B $1450
   C $1500
   D $2000

10. If the Stewarts can get a 5-year loan with 5.75% simple interest, which of the loans is the best deal?
    F 4 year, 6%
    G 5 year, 5.75%
    H 5 year, 6.25%
    J Cannot be determined

Name _____ Date _____ Class _____

LESSON 6-7
Reading Strategies
Focus on Vocabulary

Interest is the amount of money the bank pays you to use your money, or the amount of money you pay the bank to borrow its money.

Principal is the amount of money you save or borrow from the bank.

Rate of interest is the percent rate on money you save or borrow.

Time is the number of years the money is saved or borrowed.

Use this information to answer Exercises 1–3:
You put $800 in a savings account at 4% interest and leave it there for five years.

1. What is the principal?
   _____

2. What is the interest rate?
   _____

3. What is the amount of time the money will stay in the account?
   _____

You can find out how much interest you would earn on that money by using this formula:

Interest	=	principal	•	rate	•	time	← words
I	=	p	•	r	•	t	← symbols
I	=	$800	•	4%	•	5	
I	=	$800	•	0.04	•	5	← Change % to decimal.
I	=	$160					← Multiply to solve.

4. To find out how much interest you will earn by keeping your money in a bank, what three things do you need to know?
   _____
   _____

Holt Mathematics

Name _____ Date _____ Class _____

Puzzles, Twisters & Teasers
LESSON 6-7 Your Lucky Number!

Fill in the blanks to complete the chart.
Use the letters next to the answers to solve the riddle.

$ amount	Interest Rate	Years	Interest	Total Amount
$225	5%	3	$33.75	S
$4250	7%	1.5	L	$4696.25
$397	5%	1	R	$416.85
$700	6.25%	2	$87.50	
$775	8%	1	$62.00	O
$650	4.5%	2	E	$708.50
$2975	6%	1	I	$3153.50
$500	9%	3	$135.00	T
$1422	3%	5	G	$1635.30
$1500	3.85%	6	N	$1846.50

Why did the banker quit his job?

Because he was

___ ___ S ___ N ___
446.25 837.00   178.50   213.30

I ___ ___ ___ ___ E ___ T.
  346.50 635.00 58.50 19.85   258.75

Practice A
6-1 Relating Decimals, Fractions, and Percents

Find the missing ratio or percent equivalent for each letter on the number line.

```
0%   r    25%  40%  a   60%   m   84% x  100%
 0   1/8   b    t   1/2   d   3/4  c  9/10  1
```

1. a 50% **2.** b $\frac{1}{4}$ **3.** c $\frac{21}{25}$ **4.** d $\frac{3}{5}$

5. m 75% **6.** r 12.5% **7.** t $\frac{2}{5}$ **8.** x 90%

Compare. Write <, >, or =.

9. $\frac{1}{2}$ > 20%   **10.** 60% < $\frac{4}{5}$ **11.** 37% = 0.37

12. 0.76 < 81%   **13.** $\frac{9}{10}$ < 99%   **14.** 0.7 > 7%

Order the numbers from least to greatest.

15. $\frac{3}{4}$, 0.4, 0.34, 45%   0.34, 0.4, 45%, $\frac{3}{4}$

16. $\frac{2}{3}$, 30%, 0.03, 23%   0.03, 23%, 30%, $\frac{2}{3}$

17. 100%, 0.95, 59%, $\frac{1}{9}$ $\frac{1}{9}$, 59%, 0.95, 100%

18. 62%, $\frac{1}{2}$, 0.58, 85% $\frac{1}{2}$, 0.58, 62%, 85%

19. Katrina spent $\frac{1}{2}$ of the money she received for her birthday on a new sweater. What percent of the money she received did she spend on the sweater?   50%

20. Mr. Laschat has 20 students in his class. If 12 of them are girls, what percent of the class are girls?   60%

Practice B
6-1 Relating Decimals, Fractions, and Percents

Find the missing ratio or percent equivalent for each letter on the number line.

```
0%  a    22%  b       56% 64% 70%  d   100%
 0  6/100  m   9/25  9/20  t  c   x   4/5   1
```

1. a 6%   **2.** b 36%   **3.** c $\frac{16}{25}$ **4.** d 80%

5. m $\frac{11}{50}$ **6.** r 45%   **7.** t $\frac{14}{25}$ **8.** x $\frac{7}{10}$

Compare. Write <, >, or =.

9. $\frac{3}{4}$ > 70%   **10.** 60% = $\frac{3}{5}$ **11.** 58% < 0.6

12. 0.09 < 15%   **13.** $\frac{2}{3}$ > 59%   **14.** 0.45 > 40.5%

Order the numbers from least to greatest.

15. 99%, 0.95, $\frac{5}{9}$, 9.5%   9.5%, $\frac{5}{9}$, 0.95, 99%

16. $\frac{3}{8}$, 50%, 0.35, 38%   0.35, $\frac{3}{8}$, 38%, 50%

17. $\frac{4}{5}$, 54%, 0.45, 44.5%   44.5%, 0.45, 54%, $\frac{4}{5}$

18. $\frac{1}{3}$, 20%, 0.3, 3%   3%, 20%, 0.3, $\frac{1}{3}$

19. There are 25 students in math class. Yesterday, 6 students were absent. What percent of the students were absent?   24%

20. Albert spends 2 hours a day on his homework and an hour playing video games. What percent of the day is this?   12.5%

21. Ragu ran the first 3 miles of a 5 mile race in 24 minutes. What percent of the race has he run?   60%

Practice C
6-1 Relating Decimals, Fractions, and Percents

Compare. Write <, >, or =.

1. $\frac{1}{3}$ > 32%   **2.** 87.5% = $\frac{7}{8}$ **3.** 99% < 1

Order the numbers from least to greatest.

4. 98%, 0.94, $\frac{5}{4}$, 95.9%   0.94, 95.9%, 98%, $\frac{5}{4}$

5. $\frac{1}{5}$, 5%, 0.35, 3.5%   0.35%, 5%, $\frac{1}{5}$, 0.35

6. $\frac{5}{8}$, 65%, 0.5, 60.5%   0.5, 60.5%, $\frac{5}{8}$, 65

7. $\frac{2}{3}$, 60%, 0.06, 6.5%   0.06, 6.5%, 60%, $\frac{2}{3}$

Write the labels from each circle graph as percents.

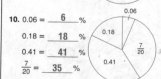

8. 0.6 = 60 %
0.1 = 10 %
0.3 = 30 %

9. $\frac{1}{6}$ = $16\frac{2}{3}$ %
$\frac{1}{4}$ = 25 %
$\frac{7}{12}$ = $58\frac{1}{3}$ %

10. 0.06 = 6 %
0.18 = 18 %
0.41 = 41 %
$\frac{7}{20}$ = 35 %

11. $\frac{1}{20}$ = 5 %
$\frac{1}{4}$ = 25 %
$\frac{11}{50}$ = 22 %
$\frac{12}{25}$ = 48 %

12. Of the 150 students in the eighth grade, 6 were absent yesterday. What percent of eighth graders were absent yesterday?   4%

13. A student has read 23 out of 41 pages assigned for homework. What percent of the pages has the student read?   56%

14. Tish has 17 relatives that live out of state and 12 that live in state. What percent of her relatives live in state?   41%

15. A molecule of nitric acid is made up of 3 molecules of oxygen, 1 molecule of hydrogen, and 1 molecule of nitrogen. What percent of the atoms of nitric acid is oxygen?   60%

Reteach
6-1 Relating Decimals, Fractions, and Percents

A *percent* (symbol %) is a *ratio*, where the comparison is to the number 100.
The ratio is then written in simplest form.

$40\% = \frac{40}{100} = \frac{40 \div 20}{100 \div 20} = \frac{2}{5}$

Write each percent as a ratio in simplest form.

1. $80\% = \frac{80}{100}$
$= \frac{80 \div 20}{100 \div 20}$
$= \frac{4}{5}$

2. $37.5\% = \frac{37.5}{100}$
$= \frac{375}{1000}$
$= \frac{375 \div 125}{1000 \div 125}$
$= \frac{3}{8}$

3. $65\% = \frac{65}{100}$
$= \frac{65 \div 5}{100 \div 5}$
$= \frac{13}{20}$

Since a percent compares a number to 100, a percent can be written as a decimal. $40\% = \frac{40}{100} = 0.40$

Write each percent as a decimal.

4. $80\% = \frac{80}{100}$
$= 0.80$

5. $37.5\% = \frac{37.5}{100}$
$= \frac{375}{1000}$
$= 0.375$

6. $65\% = \frac{65}{100}$
$= 0.65$

Use the results of Exercises 1–6 to compare. Write <, >, or =.

7. $\frac{13}{20}$ > 37.5%   **8.** 80% > 0.65

9. 65% < 0.8   **10.** 0.375 < 80%

11. 37.5% < 0.65   **12.** 65% = $\frac{13}{20}$

Holt Mathematics

LESSON 6-1 Challenge: 100% Filled

Materials needed: colored pencils or pens

For each exercise, select from the box a different combination of numbers whose sum is 100%. An item may be used only once in a combination, but may be used again in a different combination. Write your selection on the line below the grid. Shade the squares in the grid with a different color for each number you selected:

68%	$\frac{1}{2}$	60%	$\frac{1}{20}$	1%	0.9	0.04	$\frac{13}{20}$	$\frac{1}{7}$
$\frac{3}{25}$	16%	55%	0.02	$\frac{1}{9}$	0.15	$\frac{1}{8}$	24%	0.44
0.06	$\frac{1}{6}$	$\frac{1}{4}$	29%	0.037	$\frac{1}{3}$	19%	$\frac{17}{50}$	

1.
2.
3.

Possible Combinations

0.15, 60%, $\frac{1}{4}$

55%, 0.44, 1%

19%, 1%, $\frac{1}{2}$, 24%, 0.06

LESSON 6-1 Problem Solving: Relating Decimals, Fractions, and Percents

The table shows the ratio of brain weight to body size in different animals. Use the table for Exercises 1–3. Write the correct answer.

1. Complete the table to show the percent of each animal's body weight that is brain weight. Round to the nearest hundredth.

2. Which animal has a greater brain weight to body size ratio, a dog or an elephant?

   dog

3. List the animals from least to greatest brain weight to body size ratio.

   horse, elephant, dog, cat, mouse

Animal	Brain Weight / Body Weight	Percent
Mouse	$\frac{1}{40}$	2.5%
Cat	$\frac{1}{100}$	1%
Dog	$\frac{1}{125}$	0.8%
Horse	$\frac{1}{600}$	0.17%
Elephant	$\frac{1}{560}$	0.18%

The table shows the number of wins and losses of the top teams in the National Football Conference from 2004. Choose the letter of the best answer. Round to the nearest tenth.

4. What percent of games did the Green Bay Packers win?
   A 10%   C 37.5%
   B 60%   **D** 62.5%

Team	Wins	Losses
Philadelphia Eagles	13	3
Green Bay Packers	10	6
Atlanta Falcons	11	5
Seattle Seahawks	9	7

5. Which decimal is equivalent to the percent of games the Seattle Seahawks won?
   F 0.05625   H 5.625
   G 0.5625   J 56.25

6. Which team listed had the highest percentage of wins?
   A Philadelphia Eagles
   B Green Bay Packers
   C Atlanta Falcons
   D Seattle Seahawks

LESSON 6-1 Reading Strategies: Multiple Representations

Three out of four middle school students have homework each night. Three out of four is a ratio and can be shown in different ways.

The ratio can be written as a fraction. → $\frac{3}{4}$

The ratio can be written as a decimal. → 0.75
Divide the numerator by the denominator to get the equivalent decimal.
3 ÷ 4 = 0.75

The ratio can be read as a fraction with a denominator of 100. Read: → "75 hundredths"

The ratio can be written as a fraction with a denominator of 100. → $\frac{75}{100}$

Percent means "per hundred." $\frac{75}{100}$ can be written as 75%.

Use this statement to complete Exercises 1–6:
One out of five students has a younger brother.

1. Write this ratio as a fraction. — $\frac{1}{5}$
2. How can you find the decimal that is equal to $\frac{1}{5}$? — Divide 1 by 5
3. Write the decimal you get when you divide 1 by 5. — 0.2
4. What does *percent* mean? — per hundred
5. Write a fraction equal to $\frac{1}{5}$ and 0.2 that has a denominator of 100. — $\frac{20}{100}$
6. Write that fraction as a percent. — 20%

LESSON 6-1 Puzzles, Twisters & Teasers: A Slice of Soccer Fun!

To find the answer to the riddle, label the unidentified portion of each circle graph. Express the answers as decimals. Each decimal is paired with a letter. Put the letter that matches the decimal in the puzzle answer to solve the riddle.

1. N — 0.25
2. Y — 0.17
3. E — 0.05
4. S — 0.45
5. E — 0.05
6. E — 0.05
7. W — 0.10
8. J — 0.50
9. R — 0.125

Where is the best place to buy a new soccer shirt?

N   E   W   J   E   R   S   E   Y
0.25 0.05 0.10   0.50 0.05 0.125 0.45 0.05 0.17

LESSON 6-2 Practice A
Estimate with Percents

Write a benchmark you could use to estimate each percent.
Estimates may vary. Possible estimates are shown.

1. 9%   __10%__
2. 73%   __75%__
3. 26%   __25%__
4. 48%   __50%__
5. 19%   __20%__
6. 53%   __50%__
7. 34%   __$33.\overline{3}\%$ or $\frac{1}{3}$__
8. 12%   __10%__
9. 65%   __$66.\overline{6}\%$ or $\frac{2}{3}$__

Estimate.

10. 10% of 98   __about 10__
11. 48% of 83   __about 40__
12. 20% of 42   __about 8__
13. 40 out of 125   __about 30%__
14. 33% of 45   __about 15__
15. 11 out of 215   __about 5%__
16. 25% of 33   __about 8__
17. 21% of 49   __about 10__
18. 30% of 36   __about 12__

19. Felicia spent $8.78 of her $10.00 weekly allowance. About what percent of her allowance did she spend?   __about 90%__

20. A restaurant bill for dinner is a total of $49.60. Estimate the amount to leave as a 20% tip.   __about $10__

21. A company has found that on average about 8% of the light bulbs they manufacture are defective. Out of 998 bulbs, the manager assumes that about 80 are defective. Is the manager's estimate reasonable? Explain.   __Yes; 8% of 998 is about 10% of 1,000, and 10% of 1,000 is 100. Since 80 is close to 100, the estimate is reasonable.__

LESSON 6-2 Practice B
Estimate with Percents

Estimate. Estimates may vary.

1. 74% of 99   __about 75__
2. 25% of 39   __about 10__
3. 52% of 10   __about 5__
4. 21% of 50   __about 10__
5. 30% of 61   __about 20__
6. 24% of 48   __about 12__
7. 5% of 41   __about 2__
8. 50% of 178   __about 90__
9. 33% out of 62   __about 20__

Estimate.

10. 48% of 30 is about what number?   __about 15__
11. 26% of 36 is about what number?   __about 9__
12. 30% of 22 is about what number?   __about 7__
13. 21% of 63 is about what number?   __about 12__

14. Rodney's weekly gross pay is $91. He must pay about 32% in taxes and deductions. Estimate Rodney's weekly take-home pay after deductions.   __about $60__

15. In the last school election, 492 students voted. Mary received 48% of the votes. About how many votes did she receive?   __about 250__

16. A restaurant bill for lunch is $14.10. Grace wants to leave a 15% tip and the sales tax rate is 5.5%. About how much will lunch cost Grace in all?   __about $17.10__

17. A company has found that on average about 6% of the batteries they manufacture are defective. Out of 1,385 batteries, the supervisor assumes that about 83 are defective. Estimate to determine if the manager's number is reasonable? Explain.   __Yes; 6% of 1,385 is about 5% of 1,400; 10% of 1,400 is 140, and half of 140 is 70. Since 5% of 1,400 is 83 and 70 is close to 70, the estimate is reasonable.__

LESSON 6-2 Practice C
Estimate with Percents

Estimate each number of percent.

1. 74% of 59 is about what number?   __about 45__
2. 32% of 449 is about what number?   __about 150__
3. 19% of 60 is about what number?   __about 12__
4. 31% of 67 is about what number?   __about 22__
5. 9.8% of 710 is about what number?   __about 71__
6. 103% of 42 is about what number?   __about 42__
7. 39% of 78 is about what number?   __about 32__
8. 90% of 388 is about what number?   __about 360__
9. 66% of 24 is about what number?   __about 16__
10. 18.7% of 98 is about what number?   __about 20__
11. 50% of 18.25 is about what number?   __about 9__
12. $33\frac{1}{3}$% of 62 is about what number?   __about 20__

13. A stock on the New York State Stock Exchange opens at $45.02 and closes up 10% for the day. About how much does stock price increase?   __about $4.50__

14. A driver delivered 462 boxes to a manufacturing company. She unloaded 75% of the boxes before lunch. About how many boxes did she unload?   __about 300 boxes__

15. A bill for dinner is $35.75. Josh wants to leave a 15% tip. The bill will also include sales tax at a rate of 6.2%. About how much will dinner cost Josh in all?   __about $42.75__

16. On a weekday, about 39% of commuter cars have a single occupant. Out of 2,385 commuter cars, the traffic control officer assumes that about 930 cars have a single occupant. Estimate to determine if the traffic officer's number is reasonable? Explain.   __Yes; 39% of 2,385 is about 40% of 2,400; 10% of 2,400 is 240 and 4 × 240 = 960. Since 40% of 2,400 is 960 and 930 is close to 960, the estimate is reasonable.__

LESSON 6-2 Reteach
Estimate with Percents

You can estimate the solutions to different types of problems involving percents by rounding numbers.

Type 1: Finding a percent of a number.

Estimate 38% of 470.

First, round the percent to a common percent with an easy fractional equivalent.

$$38\% \approx 40\% = \frac{40}{100} = \frac{2}{5}$$

Then, round the number to a number divisible by the denominator of the fraction.   500 is divisible by 5.

$470 \approx 500$

Use mental math to multiply.

$\frac{2}{5} \cdot 500$ Think: If $500 \div 5 = 100$, then $100 \cdot 2 = 200$.

So, 38% of 470 is about 200.

Complete to estimate each percent. Estimates may vary.

1. 32% of 872

   Round to a percent with an easy fractional equivalent: $32\% \approx$ __$33\frac{1}{3}$__ % = __$\frac{1}{3}$__

   Round to hundreds place, divisible by 3: $872 \approx$ __900__

   Do mental math to multiply: $\frac{1}{3} \times$ __900__ = __300__

   So, 32% of 872 is about __300__.

2. 78% of 495

   % to easy fraction:
   $78\% \approx$ __80__ % = __$\frac{4}{5}$__

   divisible by denominator:
   $495 \approx$ __500__

   Do mental math to multiply.
   $\frac{4}{5} \times$ __500__ = __400__

   So, 78% of 495 is about __400__.

3. 73% of 1175

   % to easy fraction:
   $73\% \approx$ __75__ % = __$\frac{3}{4}$__

   divisible by denominator:
   $1175 \approx$ __1200__

   Do mental math to multiply.
   $\frac{3}{4} \times$ __1200__ = __900__

   So, 73% of 1175 is about __900__.

LESSON 6-2 Reteach
Estimate with Percents (continued)

Type 2: Finding what percent one number is of another.

About what percent of 82 is 39?

Round to numbers with common factors, **compatible numbers**.
82 ≈ 80   39 ≈ 40

Write a fraction in the form $\frac{is\ number}{of\ number}$: $\frac{40}{80}$.

Reduce the fraction and convert to percent.

$\frac{40}{80} = \frac{1}{2}$ You may recognize the % now.

$= \frac{1 \cdot 50}{2 \cdot 50} = \frac{50}{100}$ Or, get a fraction with denominator 100.

$= 50\%$

So, 39 is about 50% of 82.

Estimate what percent one number is of another. Estimates may vary.

4. About what percent of 98 is 27?

   Use compatible numbers. 98 ≈ __100__   27 ≈ __25__

   Write fraction. $\frac{is\ number}{of\ number}$: $\frac{25}{100}$

   Reduce the fraction and convert to percent. $\frac{25}{100} = \frac{1}{4} = $ __25__ %

   So, 27 is about __25__ % of 98.

5. About what percent of 19 is 6?

   19 ≈ __20__   6 ≈ __8__

   $\frac{is\ number}{of\ number}$: $\frac{8}{20} = \frac{2}{5} = $ __40__ %

   So, 6 is about __40__ % of 19.

6. About what percent of 51 is 89?

   51 ≈ __50__   89 ≈ 90

   $\frac{is\ number}{of\ number}$: $\frac{90}{50} = \frac{9}{5} = 1 + \frac{4}{5}$

   = __180__ %

   So, 89 is about __180__ % of 51.

LESSON 6-2 Challenge
As the Wheel Turns

For each wheel, write a percent problem using estimation. In your problem, use the percent at the center of the wheel and two other numbers on the wheel. The first one in each row is done.

1.

63% of 150 is about 90.

2.

68% of 90 is about 60.

3.

51% of 62 is about 33.

4.

16 is about 23% of 71.

5.

4 is about 11% of 36.

6.

21 is about 38% of 53.

7.

141% of 85 is about 120.

8.

129% of 6 is about 8.

9.

247% of 31 is about 75.

LESSON 6-2 Problem Solving
Estimate with Percents

Write an estimate.

1. A store requires you to pay 15% up front on special orders. If you plan to special order items worth $74.86, estimate how much you will have to pay up front.

   Possible answer: $11

2. A store is offering 25% off of everything in the store. Estimate how much you will save on a jacket that is normally $58.99.

   Possible answer: $15

3. A certain kind of investment requires that you pay a 10% penalty on any money you remove from the investment in the first 7 years. If you take $228 out of the investment, estimate how much of a penalty you will have to pay.

   Possible answer: $25

4. John notices that about 18% of the earnings from his job go to taxes. If he works 14 hours at $6.25 an hour, about how much of his check will go for taxes?

   Possible answer: $18

Choose the letter for the best estimate.

5. In its first week, an infant can lose up to 10% of its body weight in the first few days of life. Which is a good estimate of how many ounces a 5 lb 13 oz baby might lose in the first week of life?
   A 0.6 oz   C 18 oz
   B 9 oz   D 22 oz

6. A CD on sale costs $12.89. Sales tax is 4.75%. Which is the best estimate of the total cost of the CD?
   F $13.30   H $14.20
   G $13.55   J $14.50

7. In a class election, Pedro received 52% of the votes. There were 274 students who voted in the election. Which is the best estimate of the number of students who voted for Pedro?
   A 70 students   C 125 students
   B 100 students   **D** 140 students

8. Mel's family went out for breakfast. The bill was $24.25 plus 5.2% sales tax. Mel wants to leave a 20% tip. Which is the best estimate of the total bill?
   F $25.45   **H** $30.25
   G $29.25   J $32.25

LESSON 6-2 Reading Strategies
Use Context

An **estimate** is useful when an exact answer is not needed. You can

Chris made 15 out of 32 free throws. About what fraction of his free throws did he make?

$\frac{15}{32}$ stands for the number of free throws Chris made out of the total number of baskets shot.

30 is close to 32.

$\frac{15}{30} = \frac{1}{2}$

Chris made about half of his free throws.

There are 36 students in the eighth grade. 28% of them play basketball. **About** how many students play basketball?

25% is close to 28% and much easier to work with.

25% = one-fourth

One-fourth of 36 = 9.

use numbers that work well together when you estimate.

Use the following problem for Exercises 1–4:
27.5% of the 40 students on the field trip rode the bus. About how many students rode the bus?

1. Do 27.5 and 40 work well together?
   no

2. What percent would work better with 40? Explain why.
   25%, because 25% can be written as $\frac{1}{4}$

3. Rewrite the problem using numbers that work well together.
   $\frac{1}{4}$ of 40, or $\frac{1}{4} \cdot 40$

4. About how many students rode the bus?
   about 10 students

LESSON 6-2 Puzzles, Twisters & Teasers
Dressed to Fill!

Decide if the statements are true or false. After you have completed all of the exercises take the letters from all of the true statements and unscramble them to solve the riddle.

1. 50% of 297 is about 150 — G — true
2. 60% of 800 is about 300 — M — false
3. 50% of 795 is about 200 — A — false
4. 24% of 78 is about 20 — N — true
5. 26% of 98 is about 75 — W — false
6. 25 % of 200 is about 100 — Y — false
7. 31% of 42 is about 13 — I — true
8. 25% of 925 is about 230 — S — true
9. 88% of 180 is about 160 — E — true
10. 15% of 600 is about 150 — B — false
11. 105% of 776 is about 815 — L — true
12. 11% of 61 is about 7 — R — true
13. 55% of 810 is about 450 — D — true

Why did the tomato blush?

Because it saw the S A L A D D R E S S I N G

LESSON 6-3 Practice A
Finding Percents

Write an equation or proportion that you can use to solve each problem.

1. What percent of 14 is 7?
   Sample: $p \cdot 14 = 7$

2. 8 is what percent of 25?
   Sample: $\frac{n}{100} = \frac{8}{25}$

Find each percent.

3. What percent of 60 is 15? **25%**
4. 11 is what percent of 44? **25%**
5. What percent of 35 is 7? **20%**
6. What percent of 90 is 9? **10%**
7. 6 is what percent of 40? **15%**
8. What percent of 8 is 6? **75%**
9. What percent of 50 is 15? **30%**
10. 14 is what percent of 56? **25%**
11. 9 is what percent of 36? **25%**
12. 10 is what percent of 25? **40%**
13. What percent of 50 is 12? **24%**
14. What percent of 24 is 3? **12.5%**

15. A survey asked 180 people if they like winter, and 45 people said yes. What percent of the people surveyed like winter? **25%**

16. The Reed family bought a case of apples. Paula ate $\frac{3}{10}$ of the apples. Kyle ate 21% of the apples. Wendy ate 0.34 of the apples, and Ron ate the rest. What percent of the apples did Ron eat? **15%**

17. There are 9 sophomores on the basketball team. This is 300% of the number of freshmen on the team. Find the number of freshmen on the team. **3**

18. The Eastside Trail is 20 miles long. This is 140% as long as the Westside Trail. Find the length of the Westside Trail to the nearest mile. **14 mi**

LESSON 6-3 Practice B
Finding Percents

Find each percent.

1. What percent of 84 is 21? **25%**
2. 24 is what percent of 60? **40%**
3. What percent of 150 is 75? **50%**
4. What percent of 80 is 68? **85%**
5. 36 is what percent of 80? **45%**
6. What percent of 88 is 33? **37.5%**
7. 19 is what percent of 95? **20%**
8. 28.8 is what percent of 120? **24%**
9. What percent of 56 is 49? **87.5%**
10. What percent of 102 is 17? **$16\frac{2}{3}$**
11. What percent of 94 is 42.3? **45%**
12. 90 is what percent of 75? **120%**

13. Daphne bought a used car for $9200. She made a down payment of $1840. Find the percent of the purchase price that is the down payment. **20%**

14. Tricia read $\frac{1}{4}$ of her book on Monday. On Tuesday, she read 36% of the book. On Wednesday, she read 0.27 of the book. She finished the book on Thursday. What percent of the book did she read on Thursday? **12%**

15. An airplane traveled from Boston to Las Vegas making a stop in St. Louis. The plane traveled 2410 miles altogether, which is 230% of the distance from Boston to St. Louis. Find the distance from Boston to St. Louis to the nearest mile. **1048 mi**

16. The first social studies test had 16 questions. The second test had 220% as many questions as the first test. Find the number of questions on the second test. **36 questions**

LESSON 6-3 Practice C
Finding Percents

Find each number.

1. What number is 80% of 200? **160**
2. What number is $33\frac{1}{3}$% of 243? **81**
3. What number is 3.5% of 240? **8.4**
4. What number is 265% of 80? **212**

Complete each statement.

5. Since 12 is 15% of 80,
   24 is **30** % of 80.
   36 is **45** % of 80.
   48 is **60** % of 80.

6. Since 15 is 12% of 125,
   30 is **24** % of 125.
   45 is **36** % of 125.
   60 is **48** % of 125.

7. Since 81 is 150% of 54,
   81 is **100** % of 81.
   81 is **75** % of 108.
   81 is **25** % of 324.

8. Since 18 is 5% of 360,
   18 is **10** % of 180.
   18 is **20** % of 90.
   18 is **40** % of 45.

9. Jerry collected 130 souvenirs from 40 states in the United States and 120 from 20 other countries. What percent of the Jerry's souvenirs are from the United States? **52%**

10. Dan completed $\frac{3}{8}$ of the science poster. Becky completed 24.5% of the poster, and Greg completed 0.17 of the poster. Amber completed the rest of the poster. What percent of the poster did Amber complete? **21%**

11. The longest river in the United States is the Missouri River. It is 2,540 miles long, which is 134% of the length of the Rio Grande River. Find the length of the Rio Grande to the nearest ten feet. **1900 ft**

LESSON 6-3 Reteach
Finding Percents

Since a percent is a ratio, problems involving percent can be solved by using a proportion. There are different possibilities for an unknown quantity in this proportion.

$$\frac{\text{symbol number}}{100} = \frac{\text{is number}}{\text{of number}}$$

Possibility 1: Find the *symbol number*.

What percent of 80 is 16?

$$\frac{\text{symbol number}}{100} = \frac{\text{is number}}{\text{of number}}$$
$$\frac{x}{100} = \frac{16}{80}$$
$$80 \cdot x = 16 \cdot 100$$
$$\frac{80x}{80} = \frac{1600}{80}$$
$$x = 20 \quad \text{So, 16 is 20% of 80.}$$

Find what percent one number is of another.

1. What percent of 64 is 16?
$$\frac{x}{100} = \frac{16}{64}$$
$$64 \cdot x = 16 \cdot 100$$
$$\frac{64x}{64} = \frac{1600}{64}$$
$$x = \underline{25}$$
So, 16 is __25%__ of 64.

2. What percent of 200 is 150?
$$\frac{x}{100} = \frac{150}{200}$$
$$200 \cdot x = 150 \cdot 100$$
$$\frac{200x}{200} = \frac{15{,}000}{200}$$
$$x = \underline{75}$$
So, 150 is __75%__ of 200.

3. What percent of 4 is 6?
$$\frac{x}{100} = \frac{6}{4}$$
$$4 \cdot x = 6 \cdot 100$$
$$\frac{4x}{4} = \frac{600}{4}$$
$$x = \underline{150}$$
So, 6 is __150%__ of 4.

4. About what percent of 115 is 40?
$$\frac{x}{100} = \frac{40}{115}$$
$$115 \cdot x = 40 \cdot 100$$
$$\frac{115x}{115} = \frac{4000}{115}$$
$$x = \underline{34.7}$$
So, 40 is about __35%__ of 115.

LESSON 6-3 Reteach
Finding Percents (continued)

Possibility 2: Find the *is number*.

What is 20% of 80?

$$\frac{\text{symbol number}}{100} = \frac{\text{is number}}{\text{of number}}$$
$$\frac{20}{100} = \frac{x}{80}$$
$$100 \cdot x = 20 \cdot 80$$
$$\frac{100x}{100} = \frac{1600}{100}$$
$$x = 16 \quad \text{So, 20% of 80 is 16.}$$

Find the indicated percent of each number.

5. What is 30% of 150?
$$\frac{30}{100} = \frac{x}{150}$$
$$\underline{100} \cdot x = \underline{30} \cdot \underline{150}$$
$$\frac{100x}{100} = \frac{4500}{100}$$
$$x = \underline{45}$$
So, 30% of 150 is __45__.

6. What is 75% of 205?
$$\frac{75}{100} = \frac{x}{205}$$
$$\underline{100} \cdot x = \underline{75} \cdot \underline{205}$$
$$\frac{100x}{100} = \frac{15{,}375}{100}$$
$$x = \underline{153.75}$$
So, 75% of 205 is __153.75__.

7. What is 125% of 300?
$$\frac{125}{100} = \frac{x}{300}$$
$$\underline{100} \cdot x = \underline{125} \cdot \underline{300}$$
$$\frac{100x}{100} = \frac{37{,}500}{100}$$
$$x = \underline{375}$$
So, 125% of 300 is __375__.

8. What is $66\frac{2}{3}$% of 81?
$$\frac{66\frac{2}{3}}{100} = \frac{x}{81}$$
$$\underline{100} \cdot x = 66\frac{2}{3} \cdot 81$$
$$\underline{100} \cdot x = \frac{200}{3} \cdot 81$$
$$\underline{100} \cdot x = \underline{5400}$$
$$\frac{100x}{100} = \frac{5400}{100}$$
$$x = \underline{54}$$

LESSON 6-3 Challenge
Particular Percentages

The shaded portion of this rectangle represents $\frac{1}{4}$, or 25%, of the whole rectangle.

Find the percent of each figure that is shaded. Lines intersecting sides or circumferences divide them equally.

1.
12.5%

2.
37.5%

3.
33.33%

4.
60%

5.
6.25%

6.
16.67%

7.
50%

8.
50%

9.
25%

10.
6.25%

11.
25%

12.
25%

LESSON 6-3 Problem Solving
Finding Percents

Write the correct answer.

1. Florida State University in Tallahassee, Florida has 29,820 students. Approximately 60% of the students are women. How many of the students are women?

   __17,892 students__

2. The yearly cost of tuition, room and board at Florida State University for a Florida resident is $10,064. If tuition is $3,208 a year, what percent of the yearly cost is tuition? Round to the nearest tenth of a percent.

   __31.9%__

3. The yearly cost of tuition, room and board at Florida State University for a non-Florida resident is $23,196. If tuition is $16,340 a year, what percent of the yearly cost is tuition for a non-resident? Round to the nearest tenth of a percent.

   __70.4%__

4. Approximately 65% of the students who apply to Florida State University are accepted. If 15,000 students apply to Florida State University, how many would you expect to be accepted?

   __9750 students__

The top four NBA field goal shooters for the 2003–2004 regular season are given in the table below. Choose the letter for the best answer.

Player	Attempts	Made	Percent
Shaquille O'Neal	948	554	
Donyell Marshall	604		56.6
Elton Brand	950	348	
Dale Davis	913		53.5

NBA Field Goal Leaders 2003–2004 Season

5. What percent of field goals did Shaquille O'Neal make? Round to the nearest tenth of a percent.
   A 0.6% (C) 58.4%
   B 1.71% D 59.2%

6. How many field goals did Donyell Marshall make in the 2003–2004 regular season?
   F 38   H 295
   G 114 (J) 342

7. What percent of field goals did Elton Brand make? Round to the nearest tenth of a percent.
   A 1.87% C 51.9%
   B 50%   (D) 36.6%

8. How many field goals did Dale Davis make in the 2003–2004 regular season?
   F 274   H 457
   G 378   (J) 488

Holt Mathematics

LESSON 6-3 Reading Strategies
Read a Diagram

You can use this information to help you solve percent problems.
- When something is unknown, it is shown as a letter.
- The word *of* means to multiply.
- The word *is* means "is equal to."

The diagrams below show you how to set up two different types of percent problems.

Finding the Percent One Number Is of Another Number

What percent	of 150	is 60?
↓	↓	↓
unknown	times 150	equals 60
↓	↓	↓
p	•150	= 60

Finding a Percent of a Number

What number	is 25%	of 32?
↓	↓	↓
unknown	equals 25 percent	times 32
↓	↓	↓
s	= 25%	• 32

Answer each question.

1. What word means "to multiply"?
   of

2. What symbol is used to stand for a piece of information that is not known?
   a letter

Rewrite the percent problems using symbols.

3. What percent of 70 is 35?
   $p • 70 = 35$

4. 12 is what percent of 120?
   $12 = p • 120$

LESSON 6-3 Puzzles, Twisters & Teasers
Dino-math!

Fill in the blanks to complete each statement. Then use your answers to match the letter to the riddle.

1. 18 is __30__ % of 60   Y
2. 8 is __10__ % of 80   W
3. 20 is __50__ % of 40   D
4. 12.5 is __1__ % of 1250   A
5. 150 is __187.5__ % of 80   N
6. 35 is __49.3__ % of 71   S
7. 6 is __13.6__ % of 44   V
8. 25 is __16.7__ % of 150   N
9. 74 is __33.3__ % of 222   I
10. 225 is __19.9__ % of 1130   T

Why did the dinosaur cross the road?

Because the chicken W A S N ' T
                    10 1 49.3 16.7 19.9

I N V E N T E D Y E T
33.3   13.6   187.5   50   30

LESSON 6-4 Practice A
Finding a Number When the Percent Is Known

Write an equation or proportion that you can use to find each number.

1. 9 is 15% of what number?
   Sample: $9 = 0.15n$

2. 80% of what number is 20?
   Sample: $\frac{80}{100} = \frac{20}{n}$

Find each number.

3. 16 is 64% of what number? 25
4. 50% of what number is 12? 24
5. 3 is 5% of what number? 60
6. 7 is 35% of what number? 20
7. 75% of what number is 15? 20
8. 8 is 25% of what number? 32
9. 60% of what number is 18? 30
10. 16 is 40% of what number? 40
11. 7 is 17.5% of what number? 40
12. 25% of what number is 9.25? 37
13. 24 is 120% of what number? 20
14. $33\frac{1}{3}$ % of what number is 13? 39

15. At football practice Ty made 24 field goals, which was 80% of the field goals he attempted. How many field goals did he attempt?   30 attempts

16. At Washington Heights School there are 84 eighth graders, which is 8% of the school's enrollment. How many students attend Washington Heights School?   1050 students

17. Jackie spends 25% of her monthly salary on rent. Her rent is $650 a month. What is Jackie's monthly salary? $2600

18. Lisa's starting pay was $8.50 per hour. After 8 months she was given a 6% increase. How much per hour was Lisa's raise? $0.51

LESSON 6-4 Practice B
Finding a Number When the Percent Is Known

Find each number to the nearest tenth.

1. 40% of what number is 18? 45
2. 28 is 35% of what number? 80
3. 21 is 60% of what number? 35
4. 25% of what number is 19? 76
5. 40% of what number is 22? 55
6. 41 is 50% of what number? 82
7. 50 is 15% of what number? 333.3
8. 0.3% of what number is 24? 8,000
9. 36 is 30% of what number? 120
10. 26 is 75% of what number? 34.7
11. 12.5% of what number is 14? 112
12. 25% of what number is 28.25? 113
13. 27 is $33\frac{1}{3}$% of what number? 81
14. 54 is 150% of what number? 36

15. There were 546 students at a school assembly. This was 65% of all students who attend Content Middle School. How many students attend Content Middle School?
    840 students

16. On his last test Greg answered 64 questions correctly. This was 80% of the questions. How many questions were on the test?
    80 questions

17. The price of a jacket is $48. If the sales tax rate is 5.5%, what is the amount of tax? What is the total cost of the jacket?
    tax is $2.64; total cost is $50.64

18. Carla has finished swimming 14 laps in swim practice. This is 70% of the total number of laps she must swim. How many more laps must Carla swim to complete her practice?
    6 laps

LESSON 6-4 Practice C
Finding a Number When the Percent Is Known

Find each number to the nearest tenth.

1. 65% of what number is 78?
   120

2. 112 is 75% of what number?
   149.3

3. 175% of what number is 70?
   40

4. 46 is 2% of what number?
   2300

5. 8 is 125% of what number?
   6.4

6. 0.25% of what number is 54?
   21,600

Complete each statement

7. Since 3% of 400 is 12,
   6% of **200** is 12.
   12% of **100** is 12.
   24% of **50** is 12.

8. Since 100% of 32 is 32,
   80% of **40** is 32.
   50% of **64** is 32.
   25% of **128** is 32.

9. Since 15% of 600 is 90,
   30% of **300** is 90.
   60% of **150** is 90.
   120% of **75** is 90.

10. Since 40% of 60 is 24,
    30% of **80** is 24.
    20% of **120** is 24.
    10% of **240** is 24.

11. Parker bought a suit on sale for $162. He paid 72% of the regular price. What was the regular price of the suit?
    $225

12. Ricardo and Tyler went out for dinner. Tyler's dinner cost $18.50 and Ricardo's dinner cost $17.50. They must pay 6% tax on the meals. The two men also want to leave a 20% tip. They decide to divide the bill evenly. How much will each pay? (Hint: do not pay tax on the tip.)
    $22.68

LESSON 6-4 Reteach
Finding a Number When the Percent Is Known

Since a percent is a ratio, problems involving percent can be solved by using a proportion.

To find a number when the percent is known, the variable appears in the *of* position in the proportion

$$\frac{\text{symbol number}}{100} = \frac{\text{is number}}{\text{of number}}$$

16 is 20% of what number?
$\frac{20}{100} = \frac{16}{x}$
$20 \cdot x = 16 \cdot 100$
$\frac{20x}{20} = \frac{1600}{20}$
$x = 80$
So, 16 is 20% of 80.

Find each number whose percentage is given.

1. 18 is 75% of what number?
   $\frac{75}{100} = \frac{18}{x}$
   $75 \cdot x = \underline{18} \cdot \underline{100}$
   $\frac{75x}{75} = \frac{1800}{75}$
   $x = \underline{24}$
   So, 18 is 75% of **24**.

2. 96 is 40% of what number?
   $\frac{40}{100} = \frac{96}{x}$
   $40 \cdot x = \underline{96} \cdot \underline{100}$
   $\frac{40x}{40} = \frac{9600}{40}$
   $x = \underline{240}$
   So, 96 is 40% of **240**.

3. 7 is 125% of what number
   $\frac{125}{100} = \frac{7}{x}$
   $125 \cdot x = \underline{7} \cdot \underline{100}$
   $\frac{125x}{125} = \frac{700}{125}$
   $x = \underline{5.6}$
   So, 7 is 125% of **5.6**.

4. 40 is about 30% of what number?
   $\frac{30}{100} = \frac{40}{x}$
   $30 \cdot x = \underline{40} \cdot \underline{100}$
   $\frac{30x}{30} = \frac{4000}{30}$
   $x = \underline{133.3}$
   So, 40 is about 30% of **133.3**.

LESSON 6-4 Challenge
In the Chemistry Laboratory

When a chemist dilutes pure acid with another substance, the resulting mixture is no longer pure acid.

Consistent with the words, *pure acid* is 100% acid.
So, there are 20 grams of pure acid in 20 grams of a pure-acid solution.

Laura, a chemist, has 20 grams of a solution that is only 40% acid.

1. How many grams of pure acid are there in Laura's acid solution?
   40% of 20 grams = 8 grams

Suppose, now, Laura wants to increase the acid content of the 40% acid solution to make it a 50%-acid solution.

2. What do you think Laura has to do to increase the acid content of the solution? **Possible answer:**
   Add some pure acid; also possible to evaporate.

Laura decides to add n ounces of pure acid to increase the acid content of the original 20 grams of 40%-acid solution to make it a 50%-acid solution.

3. Represent in terms of n the total number of grams in the new solution.
   $20 + n$

4. Represent in terms of n the number of grams of pure acid in the new solution.
   $0.50(20 + n)$

Then, the amount of pure acid in the original solution plus the amount of pure acid added equals the amount of pure acid in the new solution.

5. Use the results of Exercises 1 and 4 to write an equation that will find the number n of grams of pure acid that will be added to the original solution to increase its acid content from 40% to 50%. Solve the equation.
   $8 + n = 0.50(20 + n)$
   $8 + n = 10 + 0.50n$
   $n - 0.50n = 10 - 8$
   $0.50n = 2$
   $n = 4$

6. Explain how to check your result. **Possible answer:**
   There are 24 g in all in the new solution; 50%, or 12 g, are pure acid.
   This is consistent with adding 4 g of pure acid to the original solution
   that had 8 g of pure acid.

LESSON 6-4 Problem Solving
Finding a Number When the Percent is Known

Write the correct answer.

1. The two longest running Broadway shows are *Cats* and *A Chorus Line*. *A Chorus Line* had 6137, or about 82% of the number of performances that *Cats* had. How many performances of *Cats* were there?
   7484

2. *Titanic* and *Star Wars* have made the most money at the box office. *Star Wars* made about 76.7% of the money that *Titanic* made at the box office. If *Star Wars* made about $461 million, how much did *Titanic* make? Round to the nearest million dollars.
   $601 million

Use the table below. Round to the nearest tenth of a percent.

3. What percent of students are in Pre-K through 8th grade?
   71.2%

4. What percent of students are in grades 9–12?
   28.8%

Public Elementary and Secondary School Enrollment, 2001

Grades	Population (in thousands)
Pre-K through grade 8	33,952
Grades 9–12	13,736
Total	47,688

Choose the letter for the best answer.

5. In 2000, women earned about 72.2% of what men did. If the average woman's weekly earnings was $491 in 2000, what was the average man's weekly earnings? Round to the nearest dollar.
   A $355 **C $680**
   B $542 D $725

6. The highest elevation in North America is Mt. McKinley at 20,320 ft. The highest elevation in Australia is Mt. Kosciusko, which is about 36% of the height of Mt. McKinley. What is the highest elevation in Australia? Round to the nearest foot.
   F 5480 ft H 12,825 ft
   G 7315 ft J 56,444 ft

7. The Gulf of Mexico has an average depth of 4,874 ft. This is about 36.2% of the average depth of the Pacific Ocean. What is the average depth of the Pacific Ocean? Round to the nearest foot.
   A 1764 ft C 10,280 ft
   B 5843 ft **D 13,464 ft**

8. Karl Malone is the NBA lifetime leader in free throws. He attempted 11,703 and made 8,636. What percent did he make? Round to the nearest tenth of a percent.
   F 1.4% **H 73.8%**
   G 58.6% J 135.6%

LESSON 6-4 Reading Strategies
Connecting Words and Symbols

To write equations for percent problems, connect words with symbols. Then change percents to decimals.

28 is 10% of what number? ← words
28 = 10% · n ← symbols
28 = 0.10 · n ← decimal form

Answer each question.

1. What is the decimal form of 10%?

   0.10

2. What symbol stands for "what number"?

   n

3. What symbol stands for "of"?

   ·

75 is 50% of what number? ← words
75 = 50% · n ← symbols
75 = 0.50 · n ← decimal form

Answer each question.

4. What is the decimal form of 50%?

   0.50

5. What symbol stands for "is"?

   =

6. Write the words that mean n.

   what number

7. Write the symbols for this percent problem: 50 is 6% of what number?

   50 = 6% · n

LESSON 6-4 Puzzles, Twisters & Teasers
Your Lucky Number!

Black out all the INCORRECT statements. What you see from the pattern created will be your lucky number.

5 is 30% of 70	57 is 10% of 570	24 is 15% of 160	8 is 25% of 32	6 is 2% of 300
54 is 25% of 300	9 is 10% of 99	45 is 50% of 100	78 is 19% of 7	4 is 40% of 10
21 is 75% of 210	16 is 49% of 140	968 is 10% of 96	33 is 60% of 500	51 is 6% of 850
35 is 15% of 70	44 is 16% of 88	76 is 50% of 1050	22 is 40% of 88	77 is 25% of 308
65 is 100% of 6.5	43 is 67% of 536	7 is 15% of 734	18 is 10% of 230	97 is 30% of 323
24 is 13% of 96	74 is 30% of 300	640 is 5% of 100	10 is 10% of 1000	30 is 18% of 167
55 is 87% of 230	3 is 50% of 30	19 is 43% of 87	86 is 34% of 68	20 is 20% of 100

What is your lucky number? **7**

LESSON 6-5 Practice A
Percent Increase and Decrease

State whether each change represents an increase or decrease.

1. from 10 to 15   **increase**
2. from 16 to 12   **decrease**
3. from 8 to 14   **increase**

Find each percent increase or decrease to the nearest percent.

4. from 2 to 5   **increase 150%**
5. from 10 to 6   **decrease 40%**
6. from 12 to 18   **increase 50%**
7. from 8 to 5.6   **decrease 30%**
8. from 15 to 8   **decrease 47%**
9. from 21 to 15   **decrease 29%**
10. from 17 to 21   **increase 24%**
11. from 10 to 2   **decrease 80%**
12. from 4 to 9   **increase 125%**
13. from 7 to 11   **increase 57%**
14. from 3 to 9   **increase 200%**
15. from 12 to 5   **decrease 58%**

16. World Toys buys bicycles for $38 and sells them for $95. What is the percent of increase in the price?   **150%**

17. Jack bought a stereo on sale for $231. The original price was $385. What was the percent of decrease in price?   **40%**

18. Adams Clothing Store buys coats for $50 and sells them for $80. What percent of increase is this?   **60%**

19. Asabi's average in math for the first quarter of the school year was 75. His second quarter average was 81. What was the percent of increase in Asabi's grade?   **8%**

20. A shoe store is selling athletic shoes at 30% off the regular price. If the regular price of a pair of athletic shoes is $45, what is the sale price?   **$31.50**

LESSON 6-5 Practice B
Percent Increase and Decrease

Find each percent increase or decrease to the nearest percent.

1. from 16 to 20   **increase 25%**
2. from 30 to 24   **decrease 20%**
3. from 15 to 30   **increase 100%**
4. from 35 to 21   **decrease 40%**
5. from 40 to 46   **increase 15%**
6. from 45 to 63   **increase 40%**
7. from 18 to 26.1   **increase 45%**
8. from 24.5 to 21.56   **decrease 12%**
9. from 90 to 72   **decrease 20%**
10. from 29 to 54   **increase 86%**
11. from 42 to 92.4   **increase 120%**
12. from 38 to 33   **decrease 13%**
13. from 64 to 36.4   **decrease 43%**
14. from 78 to 136.5   **increase 75%**
15. from 89 to 32.9   **decrease 63%**

16. Mr. Havel bought a car for $2400 and sold it for $2700. What was the percent of profit for Mr. Havel in selling the car?   **12.5%**

17. A computer store buys a computer program for $24 and sells it for $91.20. What is the percent of increase in the price?   **280%**

18. A manufacturing company with 450 employees begins a new product line and must add 81 more employees. What is the percent of increase in the number of employees?   **18%**

19. Richard earns $2700 a month. He received a 3% raise. What is Richard's new annual salary?   **$33,372**

20. Marlis has 765 cards in her baseball card collection. She sells 153 of the cards. What is the percent of decrease in the number of cards in the collection?   **20%**

LESSON 6-5 Practice C
Percent Increase and Decrease

Find each percent increase or decrease to the nearest percent.

1. from 120 to 162 — **increase 35%**
2. from 84 to 47.04 — **decrease 44%**
3. from 72 to 46.8 — **decrease 35%**
4. from 90 to 189 — **increase 110%**
5. from 67 to 112 — **increase 67%**
6. from 153 to 109 — **decrease 29%**

Find each missing number.

7. originally: $300; new price: $450 — **50** % increase
8. originally: $850; new price: $1147.50 — **35** % increase
9. originally: $2500; new price: $825 — **67** % decrease
10. originally: $**2100**; new price: $840 — 60% decrease
11. originally: $200; new price: $**474** — 137% increase
12. originally: $4.20; new price: $6.09 — **45** % increase

13. Fandango Store buys a computer program for $244. It sells the computer program for $927.20. What is the percent of increase in the price? **280%**

14. Denise buys a shirt on sale for $21.08. This represents a 15% decrease in price. What was the original price of the shirt? **$24.80**

15. A storeowner purchases 40 shirts for $600. She then adds 40% to her cost and tags each shirt with the same selling price. What is the amount of profit for each shirt? **$6**

LESSON 6-5 Reteach
Percent Increase and Decrease

To find the percent increase:
- Find the amount of increase by subtracting the lesser number from the greater.
- Write a fraction: percent increase = $\frac{\text{amount of increase}}{\text{original amount}}$
- If possible, simplify the fraction.
- Rewrite the fraction as a percent.

The temperature increased from 60°F to 75°F. Find the percent of increase.

percent of increase = $\frac{75° - 60°}{60°} = \frac{15°}{60°} = \frac{1}{4} = 25\%$

Complete to find each percent increase.

1. Membership increased from 80 to 100.
   - $100 - 80$
   - $= 20$
   - $\frac{20}{80} = \frac{1}{4}$
   - $= 25$ %

2. Savings increased from $500 to $750.
   - $750 - 500$
   - $= 250$
   - $\frac{250}{500} = \frac{1}{2}$
   - $= 50$ %

Find the amount of increase.

percent increase = $\frac{\text{amount of increase}}{\text{original amount}}$

Change the fraction to a percent.

3. Price increased from $20 to $23.
   - $23 - 20 = 3$
   - $\frac{3}{20}$
   - 0.15
   - $\frac{3}{20} = 20\overline{)3.00} = $ **15** %

Find the amount of increase.

percent increase = $\frac{\text{amount of increase}}{\text{original amount}}$

Change the fraction to a percent.

LESSON 6-5 Reteach
Percent Increase and Decrease (continued)

To find the percent decrease:
- Find the amount of decrease by subtracting the lesser number from the greater.
- Write a fraction: percent decrease = $\frac{\text{amount of decrease}}{\text{original amount}}$
- If possible, simplify the fraction.
- Rewrite the fraction as a percent.

Carl's weight decreased from 175 lb to 150 lb. Find the percent of decrease.

percent of decrease = $\frac{175 - 150}{175} = \frac{25}{175} = \frac{1}{7} = 7\overline{)1.000}^{0.143} = 14.3\%$

Complete to find each percent decrease.

4. Enrollment decreased from 1000 to 950.
   - $1000 - 950$
   - $= 50$
   - $\frac{50}{1000} = \frac{5}{100}$
   - $= 5$ %

5. Temperature decreased from 75°F to 60°F.
   - $75 - 60$
   - $= 15$
   - $\frac{15}{75} = \frac{3}{15} = \frac{20}{100}$
   - $= 20$ %

Find the amount of decrease.

percent decrease = $\frac{\text{amount of decrease}}{\text{original amount}}$

Change the fraction to a percent.

6. Sale price decreased from $22 to $17.
   - $22 - 17 = 5$
   - $\frac{5}{22}$
   - $\frac{5}{22} = 22\overline{)5.000}^{0.227} = $ **22.7** %

Find the amount of decrease.

percent decrease = $\frac{\text{amount of decrease}}{\text{original amount}}$

Change the fraction to a percent.

LESSON 6-5 Challenge
The Ups and Downs of the Marketplace

Prices change. The price of a stock can change every few minutes. The price of a house changes over a longer period of time.

The *selling price* of an item is what someone is willing to pay for it. It is a good measure of market value.

Find the current value of each item.
Round your answer to the nearest cent.

1. a. Amy bought a baseball card for $12. To date, the value of the card increased by 30%, then decreased by 15%, and finally increased by 40%.

   Joe bought a baseball card for $12. To date, the value of the card decreased by 10%, then increased by 70%, and finally decreased by 5%.

   Whose card is currently worth more? by how much? Explain.

   Amy's, by $1.12

   $18.56 - $17.44 = $1.12

   b. By about what percent must the currently lesser-valued card increase to be of equal value with the greater-valued card? Round your answer to the nearest tenth of a percent.

   6.4%

2. a. Jorge's family bought a house for $125,000. To date, the value of the house increased by 5%, then decreased by 25%, and finally increased by 10%.

   Gene's family bought a house for $125,000. To date, the value of the house decreased by 5%, then increased by 15%, and finally decreased by 20%.

   Whose house is currently worth more? by how much? Explain.

   Gene's, by $968.75

   $109,250 - $108,281.25
   = $968.75

   b. By about what percent must the currently lesser-valued house increase to be of equal value with the greater-valued house? Round your answer to the nearest tenth of a percent.

   0.9%

LESSON 6-5 Problem Solving
Percent Increase and Decrease

Use the table below. Write the correct answer.

Fastest Growing Metropolitan Areas, 1990–2000

Metropolitan Area	Population 1990	Population 2000	Percent Increase
Las Vegas, NV	852,737	1,563,282	
Naples, FL	152,099	251,377	
Yuma, AZ	106,895		49.7%
McAllen-Edinburg-Mission, TX	383,545		48.5%

1. What is the percent increase in the population of Las Vegas, NV from 1990 to 2000? Round to the nearest tenth of a percent.
   83.3%

2. What is the percent increase in the population of Naples, FL from 1990 to 2000? Round to the nearest tenth of a percent.
   65.3%

3. What was the 2000 population of Yuma, AZ to the nearest whole number?
   160,022

4. What was the 2000 population of McAllen-Edinburg-Mission, TX metropolitan area to the nearest whole number?
   569,564

For exercises 5–7, round to the nearest tenth. Choose the letter for the best answer.

5. The amount of money spent on automotive advertising in 2000 was 4.4% lower than in 1999. If the 1999 spending was $1812.3 million, what was the 2000 spending?
   A $79.7 million   C $1892 million
   B $1732.6 million   D $1923.5 million

6. In 1967, a 30-second Super Bowl commercial cost $42,000. In 2000, a 30-second commercial cost $1,900,000. What was the percent increase in the cost?
   F 1.7%   H 442.4%
   G 44.2%   **J 4423.8%**

7. In 1896 Thomas Burke of the U.S. won the 100-meter dash at the Summer Olympics with a time of 12.00 seconds. In 2004, Justin Gatlin of the U.S. won with a time of 9.85 seconds. What was the percent decrease in the winning time?
   A 2.15%   C 21.8%
   B 17.9%   D 45.1%

8. In 1928 Elizabeth Robinson won the 100-meter dash with a time of 12.20 seconds. In 2004, Yuliya Nesterenko won with a time that was about 10.4% less than Robinson's winning time. What was Nesterenko's time, rounded to the nearest hundredth?
   F 9.83 seconds   H 12.16 seconds
   G 10.93 seconds   J 13.47 seconds

LESSON 6-5 Reading Strategies
Compare and Contrast

Percent can be used to describe change. It is shown as a ratio.

percent of change = $\frac{\text{amount of change}}{\text{original amount}}$

Compare the two lists. Change can either increase or decrease.

Increase	Decrease
A collector sold 15 CDs. Then she sold 25 more CDs.	Ben had a collection of 60 CDs. Now he has only 45 CDs.
Sales went up, so the ratio will show a **percent of increase**.	The CD collection went down, so the ratio will show a **percent of decrease**.
Change: 25 − 15 = 10 more CDs	Change: 60 − 45 = 15 fewer CDs
Percent of change = $\frac{10}{25}$	Percent of change = $\frac{15}{60}$
Change fraction to percent: 40%	Change fraction to percent: 25%

1. Compare percent of increase with percent of decrease. How are they the same?
   Possible answer: They are both ratios.

2. Write the ratio that stands for percent of change.
   $\frac{\text{amount of change}}{\text{original amount}}$

Write *percent of increase* or *percent of decrease* to describe each situation.

3. Sophie had $70 saved. She withdrew $15 from her savings.
   percent of decrease

4. Kate bought $50 worth of groceries. Then she bought $20 more.
   percent of increase

LESSON 6-5 Puzzles, Twisters & Teasers
Do Chickens Have Funny Bones?

Circle words from the list that you find.

Find a word that answers the riddle. Circle it and write it on the line.

percent   change   increase   decrease   ratio
amount   original   decimal   application   describe

```
(A P P L I C A T I O N) F R
M V N B (D E C R E A S E) C V
(O R I G I N A L) A S D C E R
U A Z X C V B N M K O I J I
N T (P E R C E N T) E W M B Y
T (I N C R E A S E) U I A L O
(C O R N Y) Q W E R T Y L C E
(C H A N G E) (D E S C R I B E)
```

What kind of jokes do chickens like best?
__**CORNY**__ ones

....TO GET TO THE OTHER SIDE!

LESSON 6-6 Practice A
Applications of Percents

Let c = the commission amount and write an equation to find the commission for the following. Do not solve.

1. 10% commission on $4000
   $c = 0.1 \cdot 4000$

2. 6% commission on $8450
   $c = 0.06 \cdot 8450$

3. 8% commission on $3575
   $c = 0.08 \cdot 3575$

4. 12% commission on $12,750
   $c = 0.12 \cdot 12{,}750$

5. 5.5% commission on $60,000
   $c = 0.055 \cdot 60{,}000$

6. $6\frac{1}{4}$% commission on $85,900
   $c = 0.0625 \cdot 85{,}900$

Write a proportion to represent the following. Do not solve.

7. What percent of 14 is 7?
   $\frac{n}{100} = \frac{7}{14}$

8. 7 is what percent of 25?
   $\frac{n}{100} = \frac{7}{25}$

9. What number is 12.5% of 16?
   $\frac{12.5}{100} = \frac{n}{16}$

10. 21 is 35% of what number?
    $\frac{35}{100} = \frac{21}{n}$

Solve.

11. 45 is 25% of what number?
    180

12. What percent of 288 is 36?
    12.5%

13. A financial investment broker earns 4% on each customer dollar invested. If the broker invests $50,000, what is the commission on the investment?
    $2000

14. Sharlene bought 4 CDs at the music store. Each cost $14.95. She was charged 5% sales tax on her purchase. What was the total cost of her purchase?
    $62.79

15. Isaac earned $1,800 last month. He put $270 into savings. What percent of his earnings did Isaac put in savings?
    15%

16. Edel works for a company that pays a 15% commission on her total sales. If she wants to earn $450 in commissions, how much do her total sales have to be?
    $3000

LESSON 6-6 Practice B
Applications of Percents

Complete the table to find the amount of sales tax for each sale amount to the nearest cent.

1.

Sale amount	5% sales tax	8% sales tax	6.5% sales tax
$67.50	$3.38	$5.40	$4.39
$98.75	$4.94	$7.90	$6.42
$399.79	$19.99	$31.98	$25.99
$1250.00	$62.50	$100.00	$81.25

Complete the table to find the commission for each sale amount to the nearest cent.

2.

Sale amount	6% commision	9% commision	8.5% commission
$475.00	$28.50	$42.75	$40.38
$2450.00	$147.00	$220.50	$208.25
$12,500.00	$750.00	$1125.00	$1062.50
$98,900.00	$5934.00	$8901.00	$8406.50

3. Alice earns a monthly salary of $315 plus a commission on her total sales. Last month her total sales were $9640, and she earned a total of $1182.60. What is her commission rate? __9%__

4. Phillipe works for a computer store that pays a 12% commission and no salary. What will Phillipe's weekly sales have to be for him to earn $360? __$3000__

5. The purchase price of a book is $35.85. The sales tax rate is 6.5%. How much is the sales tax to the nearest cent? What is the total cost of the book?
__sales tax is $2.33; total cost is $38.18__

6. Who made more commission this month? How much did she make? Salesperson A made 11% of $67,530. Salesperson B made 8% of $85,740.
__Salesperson A $7428.30__

7. Jon earned $38,000 last year. He paid $6,840 towards entertainment. What percent of his earnings did Jon pay in entertainment expenses? __18%__

8. The Cougars won 62% of their games. They won 93 games. How many games did they lose? __57 games__

LESSON 6-6 Practice C
Applications of Percents

Find each commission or sales tax, to the nearest cent.

1. total sales $9450
   commission rate: 8%
   __$756__

2. total sales $21,097
   sales tax rate: 5.5%
   __$1160.34__

3. total sales $1089
   sales tax rate: $6\frac{1}{8}$%
   __$66.70__

4. total sales $16,772
   commission rate: 15%
   __$2515.80__

Find the total sales, to the nearest cent.

5. commission: $41.50
   commission rate: 8%
   __$518.75__

6. commission: $263.70
   commission rate: $4\frac{1}{2}$%
   __$5860__

7. commission: $614.25
   commission rate: 6.25%
   __$9828__

8. commission: $2250
   commission rate: 15%
   __$15,000__

9. A model car has a list price of $46.20. The model is on sale at 15% off. Find the total cost to the nearest cent after a 4.5% sales tax is added to the sale price. __$41.04__

10. An item priced at $776 has a sales tax of $48.50. Find the sales tax rate expressed as a percent. __6.25%__

11. James earned $39,600 last year. He paid $11,250 towards rent and $1,422 in car payments. What percent of his earnings did Jon pay in rent and car payments? __32%__

12. Find the total monthly pay if total sales for a month were $35,450, the commission rate is 4.5%, and the weekly base salary is $175. __$2295.25__

13. Ms. Simms is paid an 8% commission on all sales. She had sales of $89,400 for the month. Ms. Harris works for a different company, and also sold $89,400 for the month but made $447 more than Ms. Simms. What is Ms. Harris' commission rate? __8.5%__

LESSON 6-6 Reteach
Applications of Percents

Salespeople often earn a **commission**, a percent of their total sales.

Find the commission on a real-estate sale of $125,000 if the commission rate is 4%.

Write the percent as a decimal and multiply.

commission rate × amount of sale = amount of commission
0.04 × $125,000 = $5000

If, in addition to the commission, the salesperson earns a salary of $1000, what is the total pay?

commission + salary = total pay
$5000 + $1000 = $6000

Complete to find each total monthly pay.

1. total monthly sales = $170,000; commission rate = 3%; salary = $1500

   amount of commission = 0.03 × $__170,000__ = $__5100__

   total pay = $__5100__ + $1500 = $__6600__

2. total monthly sales = $16,000; commission rate = 5.5%; salary = $1750

   amount of commission = __0.055__ × $__16,000__ = $__880__

   total pay = $__880__ + $__1750__ = $__2630__

A **tax** is a charge, usually a percentage, generally imposed by a government.

Sales tax is the tax on the sale of an item or service.

If the sales tax rate is 7%, find the tax on a sale of $9.49.

Write the tax rate as a decimal and multiply.

tax rate × amount of sale = amount of tax
0.07 × $9.49 = $0.6643 ≈ $0.66

Complete to find each amount of sales tax.

3. item price = $5.19; sales tax rate = 6%

   amount of sales tax = 0.06 × $__5.19__ = $__0.3114__ ≈ $__0.31__

4. item price = $250; sales tax rate = 6.75%

   amount of sales tax = __0.0675__ × $__250__ = $__16.875__ ≈ $__16.88__

LESSON 6-6 Reteach
Applications of Percents (continued)

Use a proportion to find what percent of a person's income goes to a specific expense.

Heather earned $3,200 last month. She paid $448 for transportation. To find the percent of her earnings that she put towards transportation, write a proportion.

Think: What percent of 3200 is 448?

$\frac{n}{100} = \frac{448}{3200}$ ← Set up a proportion.

Think: $\frac{part}{whole} = \frac{part}{whole}$

$3200n = 448 \times 100$ ← Find cross products.
$3200n = 44,800$ ← Simplify.
$\frac{3200n}{3200} = \frac{44,800}{3200}$ ← Divide both sides by 3200.
$n = 14$ ← Simplify.

Heather put 14% of her earnings towards transportation.

Complete each proportion to find the percent of earnings.

5. Wayne earned $3,100 last month. He paid $837 for food. What percent of his earnings went to food?

   $\frac{n}{100} = \frac{837}{3100}$

   $3100n = \underline{837} \times 100$

   $3100n = \underline{83,700}$

   $\frac{3100n}{3100} = \frac{8,3700}{3100}$

   $n = \underline{27}$

   __27%__ of Wayne's earnings went to food.

6. Leah earned $1,900 last month. She paid $304 for utilities. What percent of her earnings went to utilities?

   $\frac{n}{100} = \frac{304}{1900}$

   $\underline{1900} \times n = \underline{304} \times 100$

   $1900n = \underline{30,400}$

   $\frac{1900n}{1900} = \frac{30,400}{1900}$

   $n = \underline{16}$

   __16%__ of Leah's earnings went to utilities.

Lesson 6-6 Challenge: Shoppers' Delight

Shoppers save money by buying items on sale. The amount by which the regular price is reduced is called a **discount**.

amount of discount = discount rate × regular price
sale price = regular price − amount of discount

Find the sale price after each discount.

1. regular price = $899; discount rate = 20%
   amount of discount = **$179.80**
   sale price = **$719.20**

2. regular price = $14.99; discount rate = 15%
   amount of discount = **$2.25**
   sale price = **$12.74**

Stores may offer discounts in a variety of ways. Use $100 as the regular price for the item to write your explanations. Possible answers are given:

3. Buy one at regular price. Get a second one for half price. Explain how this is different from getting a 50% discount.
   50% discount: $50 for 1 item and $100 for 2 items
   2nd item half price: $100 for first item and $150 for 2 items

4. Buy two. Get one free. Explain how this is different from getting a $33\frac{1}{3}$% discount.
   Buy 2, get 1 free: $200 for 3 items
   $33\frac{1}{3}$% discount: $200 for 3 items

5. This item is marked down by 10%. Use a coupon and get an additional 10% off. Explain how this is different from getting a 20% discount.
   20% discount on $100 item: pay $80
   After first 10% discount, item price is $90
   Now, take 10% off $90 and final price is $81.

6. An item is marked "50% off — Today Only Get Another 50% off". Explain why the item is not free.
   The additional 50% is off 50% of the reduced price.
   The item is 25% of the original price.

Lesson 6-6 Problem Solving: Applications of Percents

Write the correct answer.

1. The sales tax rate for a community is 6.75%. If you purchase an item for $500, how much will you pay in sales tax?
   $33.75

2. A community is considering increasing the sales tax rate 0.5% to fund a new sports arena. If the tax rate is raised, how much more will you pay in sales tax on $500?
   $2.50

3. Trent earned $28,500 last year. He paid $8,265 for rent. What percent of his earnings did Trent pay for rent?
   29%

4. Julie has been offered two jobs. The first pays $400 per week. The second job pays $175 per week plus 15% commission on her sales. How much will she have to sell in order for the second job to pay as much as the first?
   $1500

Choose the letter for the best answer. Round to the nearest cent.

5. Clay earned $2,600 last month. He paid $234 for entertainment. What percent of his earnings did Clay pay in entertainment expenses?
   Ⓐ 9%
   B 11%
   C 30%
   D 90%

6. Susan's parents have offered to help her pay for a new computer. They will pay 30% and Susan will pay 70% of the cost of a new computer. Susan has saved $550 for a new computer. With her parents help, how expensive of a computer can she afford?
   F $165.00   H $1650.00
   Ⓖ $785.71   J $1833.33

7. Kellen's bill at a restaurant before tax and tip is $22.00. If tax is 5.25% and he wants to leave 15% of the bill including the tax for a tip, how much will he spend in total?
   A $22.17   **Ⓒ $26.63**
   B $26.46   D $27.82

8. The 8th grade class is trying to raise money for a field trip. They need to raise $600 and the fundraiser they have chosen will give them 20% of the amount that they sell. How much do they need to sell to raise the money for the field trip?
   F $120.00   **Ⓗ $3000.00**
   G $857.14   J $3200.00

Lesson 6-6 Reading Strategies: Focus On Vocabulary

A **commission** is a percent of money a person is paid for making a sale. Many salespeople receive a commission on the amount they sell.

The **commission rate** is the percent paid on a sale. A salesperson might receive a 5% commission in addition to his salary. The commission rate is 5%.

The formula for finding out how much a salesperson earns based on the commission rate and the amount of sales is:

commission rate • sales = amount of commission

Sales tax is added to the price of an item or service. Sales tax is a percent of the purchase price. A sales tax of 6.5% means that all taxable items will have an additional 6.5% added to the total cost.

sales tax rate • sale price = sales tax
sale price + sales tax = total sale

The **total sale** price is computed by adding the sales tax to the cost of all the items purchased.

Write *commission, commission rate, sales tax,* or *total sale* to describe each situation.

1. $5.45 was added to the price of the shoes Jill bought.
   sales tax

2. The man who sold your family a car receives $500 for the sale.
   commission

3. Mr. Adams makes a 4% commission on each house he sells.
   commission rate

4. Caroline spent $37.43 for two shirts plus tax.
   total sale

Lesson 6-6 Puzzles, Twisters & Teasers: One Cool Cat!

Circle words from the list that you find.
Find a word that answers the riddle. Circle it and write it on the line.

commission   sales   tax   earnings   rate
equation   percent   decimal   convert   multiply

```
S O U R P U S S M V C D P
A R F U Y H B C U W E E L
L M J I P L U O L O R C M
E Q U A T I O N T A X I N
S D E T Y I V I A Z M J
R A T E I O P E P N H A I
P E R C E N T R L P O U
W E R T Y U I T Y S W E B
C O M M I S S I O N V Y G
Z W I T H H O L D I N G Y
Z X C E T I J O P L Y R E
Q E A R N I N G S V N O C
```

What do you call a cat that drinks lemonade?

A **SOURPUSS**

Practice A
6-7 Simple Interest

Write the formula to compute the missing value. Do not solve.

1. principal = $100
   rate = 4%
   time = 2 years
   interest = ?

   $I = 100 \cdot 0.04 \cdot 2$

2. principal = $150
   rate = ?
   time = 2 years
   interest = $9

   $r = \dfrac{9}{150 \cdot 2}$

3. principal = $200
   rate = 5%
   time = ?
   interest = $10

   $t = \dfrac{10}{200 \cdot 0.05}$

4. principal = ?
   rate = 3%
   time = 4 years
   interest = 30

   $P = \dfrac{30}{4 \cdot 0.03}$

5. Jules borrowed $500 for 3 years at a simple interest rate of 6%. How much interest will be due at the end of 3 years? How much will Jules have to repay?

   Interest due is $90; amount to be repaid is $590.

6. Karin maintained a balance of $250 in her savings account for 8 years. The financial institution paid simple interest of 4%. What was the amount of interest earned?

   $80

Complete the table.

	Principal	Rate	Time	Interest
7.	$300	3%	4 years	**$36**
8.	$450	5%	3 years	$67.50
9.	$500	4.5%	**5 years**	$112.50
10.	**$675**	8%	2 years	$108
11.	$700	4%	3 years	**$84**
12.	$750	6%	2 years	$90
13.	$800	2.5%	**5 years**	$100

Practice B
6-7 Simple Interest

Find the missing value.

1. principal = $125
   rate = 4%
   time = 2 years
   interest = ?

   $10

2. principal = ?
   rate = 5%
   time = 4 years
   interest = $90

   $450

3. principal = $150
   rate = 6%
   time = ? years
   interest = $54

   6 years

4. principal = $200
   rate = ?%
   time = 3 years
   interest = $30

   5%

5. principal = $550
   rate = ?%
   time = 3 years
   interest = $57.75

   3.5%

6. principal = ?
   rate = $3\frac{1}{4}$%
   time = 2 years
   interest = $63.05

   $970

7. Kwang deposits money in an account that earns 5% simple interest. He earned $546 in interest 2 years later. How much did he deposit?   **$5460**

8. Simon opened a certificate of deposit with the money from his bonus check. The bank offered 4.5% interest for 3 years of deposit. Simon calculated that he would earn $87.75 interest in that time. How much did Simon deposit to open the account?   **$650**

9. Douglas borrowed $1000 from Patricia. He agreed to repay her $1150 after 3 years. What was the interest rate of the loan?   **5%**

10. What is the interest paid for a loan of $800 at 5% annual interest for 9 months?   **$30**

Practice C
6-7 Simple Interest

Find the interest and the total amount to the nearest cent.

1. $345 at 4% per year for 3 years
   interest: $41.40
   repay: $386.40

2. $782 at 3.5% per year for 4 years
   interest: $109.48
   repay: $891.48

3. $6125 at 7% per year for 2.5 years
   interest: $1071.88
   repay: $7196.88

4. $9875 at $3\frac{1}{4}$% per year for 5 years
   interest: $1604.69
   repay: $11,479.69

5. $2065 at 5.5% per year for 42 months
   interest: $397.51
   repay: $2462.51

6. $1750 at $6\frac{1}{8}$% per year for 33 months
   interest: $294.77
   repay: $2044.77

7. $900 at 11% per year for 3 months
   interest: $24.75
   repay: $924.75

8. $8417 at 18% per year for 1 month
   interest: $126.26
   repay: $8543.26

9. Will deposited $1550 in an account that pays $8\frac{3}{4}$% annually. How much would be in the account at the end of 24 months?   **$1821.25**

10. What is the annual interest rate if $7200 is invested for 15 months and earns $855 interest?   **9.5%**

11. How long will it take a deposit of $4500 at an annual rate 5.75% to earn $1035?   **4 years**

12. Noah bought a new car costing $25,350. He made a 20% down payment on the car and financed the remaining cost of the car for 5 years at 6.5%. How much interest did Noah pay on his car loan?   **$6591**

13. Mr. Silva earned $196.50 in interest in a year for an account that paid 3% interest per year. If he did not take any money out of the account during the year, how much was in the account at the start of the year?   **$6550**

Reteach
6-7 Simple Interest

Interest is money paid on an investment.
A borrower pays the interest. An investor earns the interest.

Simple interest, I, is earned when an amount of money, the *principal P*, is borrowed or invested at a *rate of interest r* for a *period of time t*.

Interest = Principal · Rate · Time
$I = P \cdot r \cdot t$

Situation 1: Find I given P, r, and t.

Calculate the simple interest on a loan of $3500 for a period of 6 months at a yearly rate of 5%.

Write the interest rate as a decimal.   5% = 0.05
Write the time period in terms of years.   6 months = 0.5 year
$I = P \cdot r \cdot t$
$I = 3500 \cdot 0.05 \cdot 0.5 = \$87.50$ ← interest earned

Find the interest in each case.

1. principal $P = \$5000$; time $t = 2$ years; interest rate $r = 6\%$

   $I = P \cdot r \cdot t = \underline{5000} \cdot 0.06 \cdot \underline{2} = \$\underline{600}$

2. principal $P = \$2500$; time $t = 3$ months; interest rate $r = 8\%$

   $I = P \cdot r \cdot t = \underline{2500} \cdot \underline{0.08} \cdot \underline{0.25} = \$\underline{50}$

Situation 2: Find t given I, P, and r.

An investment of $3000 at a yearly rate of 6.5% earned $390 in interest. Find the period of time for which the money was invested.
The investment was for 2 years.

$390 = 3000 \cdot 0.065 \cdot t$
$390 = 195t$
$\dfrac{390}{195} = \dfrac{195t}{195}$
$2 = t$

Find the time in each case.

3. $I = \$1120$; $P = \$4000$; $r = 7\%$

   $I = P \cdot r \cdot t$
   $1120 = \underline{4000} \cdot 0.07 \cdot t$
   $1120 = \underline{280}\ t$
   $\dfrac{1120}{280} = \dfrac{280t}{280}$
   $\underline{4}$ years = t

4. $I = \$812.50$; $P = \$5000$; $r = 6.5\%$

   $I = P \cdot r \cdot t$
   $812.50 = \underline{5000} \cdot \underline{0.065} \cdot t$
   $812.50 = \underline{325}\ t$
   $\dfrac{812.50}{325} = \dfrac{325t}{325}$
   $\underline{2.5}$ years = t

LESSON 6-7 Reteach
Simple Interest

Situation 3: Find r given I, P, and t.
$2500 was invested for 3 years and earned $450 in interest. Find the rate of interest.

$I = P \cdot r \cdot t$
$450 = 2500 \cdot r \cdot 3$
$450 = 7500r$
$\frac{450}{7500} = \frac{7500r}{7500}$
$0.06 = r$

The interest rate was 6%.

Find the interest rate in each case.

5. $I = \$1200$; $P = \$6000$; $t = 4$ years
$I = P \cdot r \cdot t$
$1200 = \underline{6000} \cdot r \cdot 4$
$1200 = \underline{24{,}000}\ r$
$\frac{1200}{24{,}000} = \frac{24{,}000r}{24{,}000}$
$\underline{0.05} = r$
The interest rate was $\underline{5}$ %.

6. $I = \$325$; $P = \$2000$; $t = 2.5$ years
$I = P \cdot r \cdot t$
$325 = \underline{2000} \cdot r \cdot \underline{2.5}$
$325 = \underline{5000}\ r$
$\frac{325}{5000} = \frac{5000r}{5000}$
$\underline{0.065} = r$
The interest rate was $\underline{6.5}$ %.

The total amount A of money in an account after interest has been earned, is the sum of the principal P and the interest I.

Amount = Principal + Interest
$A = P + I$

Find the amount of money in the account after $3500 has been invested for 3 years at a yearly rate of 6%.

First, find the interest earned.
$I = P \cdot r \cdot t$
$I = 3500 \cdot 0.06 \cdot 3 = \$630$ ← interest earned
Then, add the interest to the principal. $3500 + 630 = 4130$
So, the total amount in the account after 3 years is $4130.

Find the total amount in the account.

7. principal $P = \$4500$; time $t = 2.5$ years; interest rate $r = 5.5\%$
$I = P \cdot r \cdot t = \underline{4500} \cdot \underline{0.055} \cdot \underline{2.5} = \$ \underline{618.75}$
Total Amount $= P + I = 4500 + \underline{618.75} = \underline{5118.75}$
So, after 2.5 years, the total amount in the account was $\underline{5118.75}$.

LESSON 6-7 Challenge
Feather Your Nest

In these exercises, you will solve an investment problem algebraically.

Problem
Nancy invested a sum of money at 6%.
She invested a second sum, $500 more than the first, at 8%.
The total interest earned for the year was $180.
How much did Nancy invest at each rate?

1. Let x represent the sum Nancy invested at 6%. Write an expression in terms of x for the interest she earned after 1 year from the 6%-investment.
$\underline{0.06x}$

2. x represents the sum Nancy invested at 6%.
   a. Write an expression in terms of x for the sum she invested at 8%.
   $\underline{x + 500}$
   b. Write an expression in terms of x for the interest she earned after 1 year from the 8%-investment.
   $\underline{0.08(x + 500)}$

3. Using your results from Exercises 1 and 2b, write an equation in terms of x to show that the total of the interest earned from the two investments is equal to $180.
$\underline{0.06x + 0.08(x + 500) = 180}$

4. Follow these steps to solve your equation for x.
   a. Apply the Distributive Property. $\underline{0.06x + 0.08x + 40 = 180}$
   b. Collect like terms on the left side. $\underline{0.14x + 40 = 180}$
   c. Subtract. $\underline{0.14x = 140}$
   d. Divide to find x. $\underline{\frac{0.14x}{0.14} = \frac{140}{0.14}\ \ x = 1000}$

So, the sum invested at 6% is $ $\underline{1000}$
and the sum invested at 8% is $ $\underline{1500}$

5. Explain how to check your result.
$\underline{0.06(1000) = 60;\ 0.08(1500) = 120;\ 60 + 120 = 180}$

LESSON 6-7 Problem Solving
Simple Interest

Write the correct answer.

1. Joanna's parents agree to loan her the money for a car. They will loan her $5,000 for 5 years at 5% simple interest. How much will Joanna pay in interest to her parents?
$\underline{\$1250}$

2. How much money will Joanna have spent in total on her car with the loan described in exercise 1?
$\underline{\$6250}$

3. A bank offers simple interest on a certificate of deposit. Jaime invests $500 and after one year earns $40 in interest. What was the interest rate on the certificate of deposit?
$\underline{8\%}$

4. How long will Howard have to leave $5000 in the bank to earn $250 in simple interest at 2%?
$\underline{2.5\ years}$

Jan and Stewart Jones plan to borrow $20,000 for a new car. They are trying to decide whether to take out a 4-year or 5-year simple interest loan. The 4-year loan has an interest rate of 6% and the 5-year loan has an interest rate of 6.25%. Choose the letter for the best answer.

5. How much will they pay in interest on the 4-year loan?
   A $4500
   B $4800
   C $5000
   D $5200

6. How much will they repay with the 4-year loan?
   F $24,500
   G $24,800
   H $25,000
   J $25,200

7. How much will they pay in interest on the 5-year loan?
   A $5000
   B $6000
   C $6250
   D $6500

8. How much will they repay with the 5-year loan?
   F $25,000
   G $26,000
   H $26,250
   J $26,500

9. How much more interest will they pay with the 5-year loan?
   A $1000
   B $1450
   C $1500
   D $2000

10. If the Stewarts can get a 5-year loan with 5.75% simple interest, which of the loans is the best deal?
    F 4 year, 6%
    G 5 year, 5.75%
    H 5 year, 6.25%
    J Cannot be determined

LESSON 6-7 Reading Strategies
Focus on Vocabulary

Interest is the amount of money the bank pays you to use your money, or the amount of money you pay the bank to borrow its money.
Principal is the amount of money you save or borrow from the bank.
Rate of interest is the percent rate on money you save or borrow.
Time is the number of years the money is saved or borrowed.

Use this information to answer Exercises 1–3:
You put $800 in a savings account at 4% interest and leave it there for five years.

1. What is the principal?
$\underline{\$800}$

2. What is the interest rate?
$\underline{four\ percent}$

3. What is the amount of time the money will stay in the account?
$\underline{five\ years}$

You can find out how much interest you would earn on that money by using this formula:

Interest	=	principal	·	rate	·	time	← words
I	=	p	·	r	·	t	← symbols
I	=	800	·	4%	·	5	
I	=	800	·	0.04	·	5	← Change % to decimal.
I	=	160					← Multiply to solve.

4. To find out how much interest you will earn by keeping your money in a bank, what three things do you need to know?
$\underline{the\ principal,\ rate\ of\ interest,\ and\ amount\ of\ time\ you\ plan\ to\ leave\ your}$
$\underline{money\ in\ the\ bank}$

Puzzles, Twisters & Teasers
6-7 Your Lucky Number!

Fill in the blanks to complete the chart.
Use the letters next to the answers to solve the riddle.

$ amount	Interest Rate	Years	Interest	Total Amount
$225	5%	3	$33.75	$258.75 S
$4250	7%	1.5	$446.25 L	$4696.25
$397	5%	1	$19.85 R	$416.85
$700	6.25%	2	$87.50	$787.50
$775	8%	1	$62.00	$837.00 O
$650	4.5%	2	$58.50 E	$708.50
$2975	6%	1	$178.50 I	$3153.50
$500	9%	3	$135.00	$635.00 T
$1422	3%	5	$213.30 G	$1635.30
$1500	3.85%	6	$346.50 N	$1846.50

Why did the banker quit his job?

Because he was

<u>L</u> <u>O</u> S <u>I</u> N <u>G</u>
446.25 837.00   178.50   213.30

<u>I</u> <u>N</u> <u>T</u> <u>E</u> <u>R</u> E <u>S</u> T.
    346.50 635.00 58.50 19.85   258.75